Audrey L. Allison

# The ITU and Managing Satellite Orbital and Spectrum Resources in the 21st Century



INTERNATIONAL®
SPACE UNIVERSITY
isu

Springer

Audrey L. Allison
Boeing Company
Boeing, WA, USA

ISSN 2191-8171 ISSN 2191-818X (electronic)
ISBN 978-3-319-05313-4 ISBN 978-3-319-05314-1 (eBook)
DOI 10.1007/978-3-319-05314-1
Springer Cham Heidelberg New York Dordrecht London

Library of Congress Control Number: 2014934895



Printed on acid-free paper

Springer is part of Springer Science+Business Media (www.springer.com)

# SpringerBriefs in Space Development

***Series Editor:***
Joseph N. Pelton, Jr.

For further volumes:
http://www.springer.com/series/10058

This Springer book is published in collaboration with the International Space University. At its central campus in Strasbourg, France, and at various locations around the world, the ISU provides graduate-level training to the future leaders of the global space community. The university offers a 2-month Space Studies Program, a 5-week Southern Hemisphere Program, a 1-year Executive MBA, and a 1-year Master's program related to space science, space engineering, systems engineering, space policy and law, business and management, and space and society.

These programs give international graduate students and young space professionals the opportunity to learn while solving complex problems in an intercultural environment. Since its founding in 1987, the International Space University has graduated more than 3,000 students from 100 countries, creating an international network of professionals and leaders. ISU faculty and lecturers from around the world have published hundreds of books and articles on space exploration, applications, science, and development.

# Contents

# About the Author

**Audrey L. Allison** is the director of Frequency Management Services for the Boeing Company. In this role, she leads Boeing's domestic and international spectrum management in support of Boeing products, testing and operations—including licensing, compliance, strategic regulatory planning, and policy development. She also represents Boeing on a broad range of industry telecommunication policies and spectrum advocacy before the International Telecommunication Union (ITU), the Inter-American Telecommunication Commission (CITEL), and US government agencies, such as the Federal Communication Commission (FCC). In 2012, Allison was reelected by the ITU Radiocommunication Assembly to serve as a vice-chair of the ITU's Radiocommunication Advisory Group in Geneva, Switzerland for a second 4-year term. She also serves as a board member of the US ITU Association and as an alternate board member of the Satellite Industry Association. She is the chair of the FCC's World Radiocommunication Conference Advisory Committee Working Group on aeronautical, maritime, and radar issues. She previously chaired the FCC Advisory Committee's Regulatory Working Group for the three previous World Radio Conferences. Allison also serves as a member of the State Department's Advisory Committee on International Communications and Information Policy.

Allison is a member of the Adjunct Faculty of the International Space University in Strasbourg, France, and lectures on satellite regulatory issues for McGill University's Institute of Air and Space Law in Montreal. Before coming to Boeing, Allison held legal positions with the FCC, Iridium LLC, and law firms in Washington, DC. She also was a consultant on international communication policy and regulatory issues to the US Department of Defense.

Allison has a Master of Business Administration, cum laude, from the International Space University in Strasbourg, France; a Master of Law in International and Comparative Law from Georgetown University in Washington, DC; a Juris Doctor in Communications Law Studies from the Catholic University of America in Washington, DC; and a Bachelor of Arts in Communications from Pennsylvania State University.

# Abbreviations

| | |
|---|---|
| A | Advance Publication Information |
| ADM | Administration |
| API | Advance Publication Information |
| BR | Radiocommunication Bureau |
| BSS | Broadcasting Satellite Service |
| C | Coordination Request |
| CHF | Swiss Franc |
| COPUOS | Committee on Peaceful Uses of Outer Space |
| CPM | Conference Preparatory Meeting |
| CR | Coordination Request |
| EARC-63 | 1963 Extraordinary Administrative Radio Conference to Allocate Frequency Bands for Space Radiocommunication Purposes (Geneva) |
| ESA | European Space Agency |
| FCC | Federal Communications Commission |
| GSO | Geostationary-Satellite Orbit |
| HDTV | High Definition Television |
| IFIC | International Frequency Information Circular |
| ITR | International Telecommunication Regulations |
| ITU | International Telecommunication Union |
| MIFR | Master International Frequency Register |
| Mod. | Modified |
| N. | Notification |
| NASA | National Aeronautics and Space Administration |
| No. | Number |
| Not. | Notification |
| ORB-88 | 1988 World Administrative Radio Conference on the Use of Geostationary-Satellite Orbit and the Planning of Space Services Utilizing It |
| RAG | Radiocommunication Advisory Group |
| RES49 | Resolution 49 |
| Rev. | Revised |

| | |
|---|---|
| RR | Radio Regulations |
| RRB | Radio Regulation Board |
| SC-WP | Working Party of the Special Committee on Regulatory/Procedural Matters |
| UAE | United Arab Emirates |
| UK | United Kingdom of Great Britain and Northern Ireland |
| UN | United Nations |
| USA | United States of America |
| WARC | World Administrative Radiocommunication Conference |
| WARC-79 | 1979 World Administrative Radiocommunication Conference (Geneva) |
| WCIT-12 | 2012 World Conference on International Telecommunication (Dubai) |
| WRC | World Radiocommunication Conference |
| WRC-97 | 1997 World Radiocommunication Conference (Geneva) |
| WRC-2000 | 2000 World Radiocommunication Conference (Istanbul) |
| WRC-03 | 2003 World Radiocommunication Conference (Geneva) |
| WRC-07 | 2007 World Radiocommunication Conference (Geneva) |
| WRC-12 | 2012 World Radiocommunication Conference (Geneva) |
| WRC-15 | 2015 World Radiocommunication Conference (Geneva) |

# Chapter 1
# Introduction

In the beginning of the twenty-first century, we are the beneficiaries of a vast array of satellite-delivered services. These services are so embedded in our everyday lives—from navigation systems in our cars to weather maps on our phones to videos delivered to our screens—that the fact their delivery relies upon satellites is largely overlooked. The technologies needed and investment required in these space-based systems, however, remains significant. Moreover, the continuing operation of these space-based assets now depends upon the workings of a complex system of interlocking legal, regulatory, and procedural arrangements involving multiple governmental and industrial actors. It is these arrangements rooted in international law, good faith, and mutual self-interest that make it possible for the world's satellite network operators to provide vital services. These operators must rely on assured access to both the necessary radio frequency spectrum resources across their vast service areas and their position in orbit. The story behind these international arrangements that underlie the satellite infrastructure we depend upon is the subject of this book (Fig. 1.1).

As surprising as it may initially seem, the home of these complex arrangements for cooperation on spectrum and orbital slots that are so essential to the successful operation of satellite networks worldwide is an elderly intergovernmental organization based in Switzerland. The International Telecommunication Union (ITU, or the Union) was established in 1865 (originally as the International Telegraph Union) to interconnect fledgling telegraph systems as they began to span the borders of western European nations (Allison 1992; Codding and Rutkowski 1982; International Telecommunication Union 1995). The ITU is now part of the United Nations system and is charged with promoting and facilitating the extension of communications systems, including by satellite, to all the world's inhabitants. The ITU's broad mandate includes two potentially contradictory elements: facilitating equitable access to spectrum/orbital resources with regard to satellite services while promoting the advancement of technology, efficient operation, and rapid implementation of these services (International Telecommunication Union 2011; Smith 1989).

What tensions could there be between such laudable missions as equitable access on one hand and promoting the technology and efficiency on the other? To some,

A.L. Allison, *The ITU and Managing Satellite Orbital and Spectrum Resources in the 21st Century*, SpringerBriefs in Space Development,
DOI 10.1007/978-3-319-05314-1_1, 

**Fig. 1.1** A commercial telecommunications satellite orbits Earth in geostationary orbit (Image courtesy of Boeing)

equitable access means that all countries should have equal access to the orbit, meaning that equal shares of the orbital/spectrum resources should be set aside for every country in perpetuity, even if a country does not currently or foreseeably plan to deploy a satellite requiring access to these resources. Such set asides or reservations, known in ITU parlance as "planning," serve to lock in the current state of technology while leaving much of the resource underutilized. In contrast, implementing advances in technology and services as they develop enables more efficient use of the resources, such as reduced spacing between satellites and use of higher throughput satellites. Yet new technology is typically more costly and therefore less available to developing countries. Thus, a system of guaranteed, reserved slots may be fair and equitable in principle, but if slots are left fallow, or are allotted based on outmoded technological assumptions, such a system leads ultimately to provision of fewer services, less efficient use of the resources, and less promotion of advanced technology.

The focus of this book is thus on how the Union strives to achieve and maintain a balance between its dual missions of equity and efficiency. This becomes more and more difficult with the increasing saturation of the geostationary orbit by a vibrant global satellite industry, and the rising interests of developing countries in their own access to these limited resources. This ongoing global struggle was much debated at the most recent World Radiocommunication Conference in Geneva in 2012. This book traces the origins of this global challenge and reviews the various approaches the Union and its members have considered and implemented over the years, and the results of these efforts. In particular, it explores the approaches the ITU adopted to stem the overfilling of geostationary satellite network filings in order achieve and balance its dual missions. These mechanisms include the concepts of "administrative due diligence" and "satellite network cost recovery." The processes introduced in 1994 have largely been viewed as a success by ITU member states and satellite operators alike. But, despite these successes, challenges remain, and congestion continues unabated. Recent efforts to extend and strengthen these mechanisms have proven controversial, but remain on the table. Moreover, the call for equitable access by developing countries continues to be heard—loud and clear.

This book will assess the effectiveness of the ITU in the new century in managing access to spectrum and orbital resources for the world's satellite networks. It will also look to the future and the next steps in the evolution of the ITU's role in managing access to the orbit.

## References

Allison, A.: Meeting the challenges of change: the reform of the International Telecommunication Union. Fed. Comm. Law J **45**, 491 (1992)

Codding, G., Rutkowski, A.: International Telecommunication Union in a Changing World. Artech House, Norwood, MA (1982)

International Telecommunication Union: Celebrating 130 Years: 1865–1995. International Systems & Communications, London (1995)

International Telecommunication Union: Collection of the Basic Texts of the International Telecommunication Union Adopted by the Plenipotentiary Conference. International Telecommunication Union, Geneva (2011) (2011 Edition)

Smith, M.: International regulation of satellite telecommunications after the space WARC. Dissertation (D.C.L.), McGill University (1989)

## *Website*

History and background of the International Telecommunication Union: http://www.itu.int/en/history/Pages/DiscoverITUsHistory.aspx

# Chapter 2
# The Basics of Satellites and the ITU

Satellites require assured access to orbit and adequate spectrum resources in order to operate and to perform their missions. Access to these shared natural resources is managed by the ITU. This chapter provides background on the current state of the satellite industry and introduces the ITU and its role in managing access to the geostationary satellite orbit (GSO) and necessary radio frequency spectrum resources to enable successful operation of satellite networks worldwide (Fig. 2.1).

There are other satellite systems using other orbits such low Earth orbits and medium Earth orbits. These are deployed in network constellations to achieve global coverage. The ITU must also address frequency allocations for these types of systems, but the frequency and orbital issues presented by these types of systems are beyond the scope of what can be addressed in this short book. The issue of equity and efficiency, however, also arises with respect to these types of satellite systems. The ITU's largest task is to create processes and recommendations to minimize interference between satellites in GSO (where most telecommunications systems now reside) and non-GSO satellite networks.

## The Satellite Industry

From their beginnings in the late 1950s, satellites have become a significant global industry that has continued to grow even throughout the recent economic downturn. As of 2013, the global satellite industry is a 189.5—billion (US)—dollar business, comprising 62 % of the overall space industry sector.

As seen in Fig. 2.2, the industry has been growing at a steady rate of over recent years, with a 7 % growth rate in 2012, outperforming the worldwide economic growth rate of 2.3 % (Satellite Industry Association 2013 State of the Satellite Industry Report, p. 4). Of the approximately 1,000 satellites currently in operation, more than half are communications satellites, as shown in Fig. 2.3. Two-thirds of these are government satellites and the remaining one-third are commercial.

A.L. Allison, *The ITU and Managing Satellite Orbital and Spectrum Resources in the 21st Century*, SpringerBriefs in Space Development,
DOI 10.1007/978-3-319-05314-1_2, 

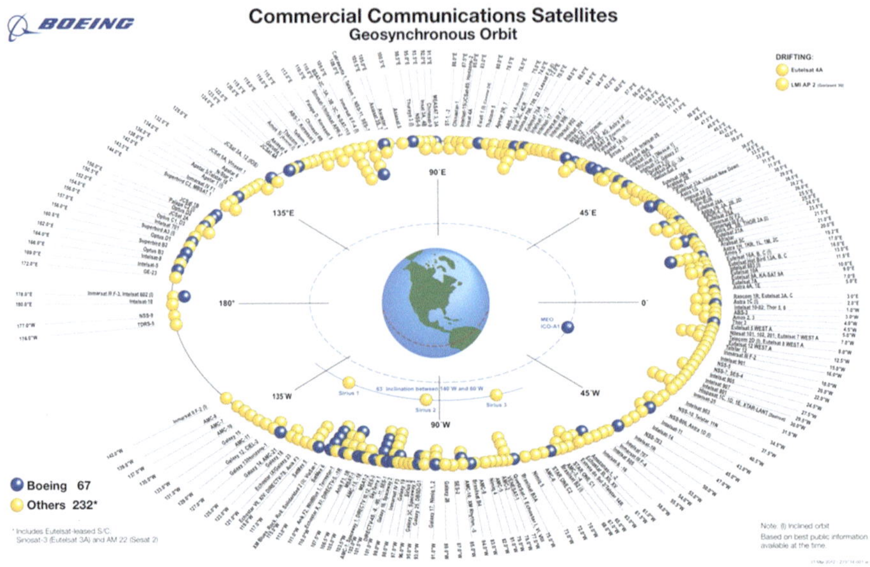


**Fig. 2.1** Commercial communications satellites in geosynchronous orbit (Image courtesy of Boeing)

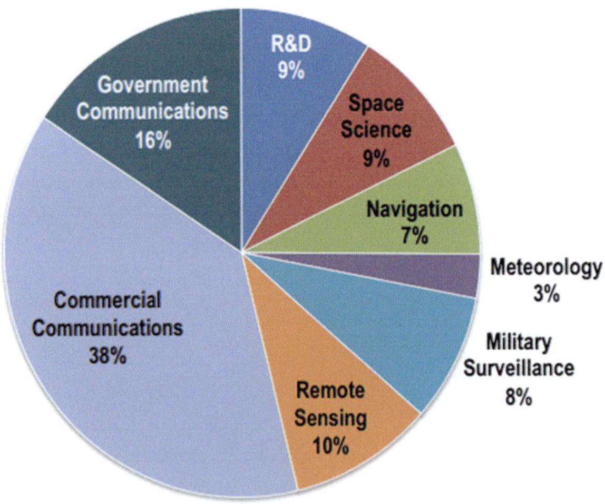


**Fig. 2.2** Operational satellites by function (Image courtesy of the Satellite Industry Association, 2014 State of the Satellite Industry Report, p. 7)

Fifty nation states are involved in the operation of these satellites, including some involved in regional operations.

All satellites, regardless of operator or type, require access to radio frequency spectra for their ability to communicate with the ground to control their operations and to fulfill their missions. In short, even satellites that are for other purposes than telecommunications, such as remote sensing, weather satellites, scientific satellites, etc.,

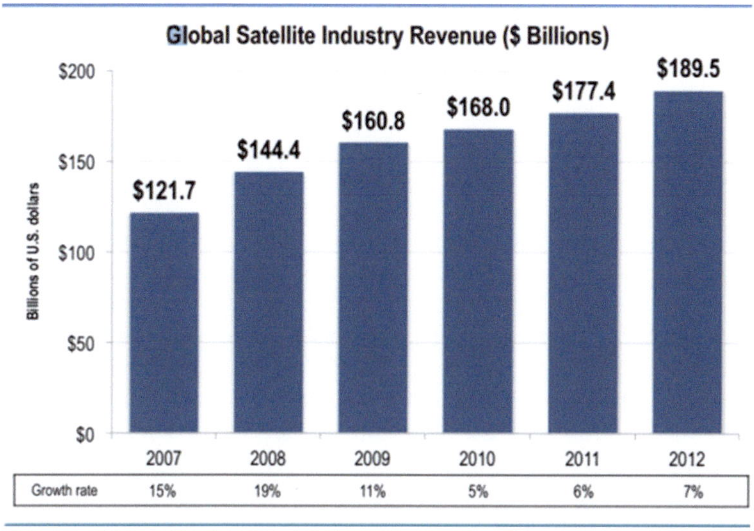


**Fig. 2.3** Global satellite industry revenues (Image courtesy of the Satellite Industry Association, 2014 State of the Satellite Industry Report, p. 4)

**Table 2.1** Satellite frequency band ranges, generally as described in ITU-R Recommendation V.431-7, 2000, p. 3 (Image courtesy of the ITU)

| Letter symbols | Radar (GHz) | | Space radio communications | |
|---|---|---|---|---|
| | Spectrum regions | Examples | Nominal designations (GHz band) | Examples (GHz) |
| L | 1–2 | 1.215–1.4 | 1.5 | 1.525–1.710 |
| S | 2–4 | 2.3–2.5<br>2.7–3.4 | 2.5 | 2.5–2.690 |
| C | 4–8 | 5.25–5.85 | 4/6 | 3.4–4.2<br>4.5–4.8<br>5.85–7.075 |
| X | 8–12 | 8.5–10.5 | – | – |
| Ku | 12–18 | 13.4–14.0<br>15.3–17.3 | 11/14<br>12/14 | 10.7–13.25<br>14.0–14.5 |
| K[a] | 18–27 | 24.05–24.25 | 20 | 17.7–20.2 |
| Ka[a] | 27–40 | 33.4–36.0 | 30 | 27.5–30.0 |
| V | – | – | 40 | 37.5–42.5<br>47.2–50.2 |

[a]For space radio communications K and Ka bands are often designated by the symbol $K_a$

must still be capable of communicating data back to Earth. This spectrum must have the necessary physical characteristics to be able to be received despite weather conditions, elevation angles, antenna characteristics, power limits, and linking over vast distances. Typically, commercial communications satellites utilize globally or regionally harmonized spectrum allocations in the so-called C, L. S. Ku, and Ka-bands, shown in Table 2.1 below. (Examples of the frequencies used in these various bands as indicated in gigahertz [or billions of cycles per second] are provided in this table.) Availability of harmonized spectrum allocations is critical because satellites typically illuminate the territory of more than a single nation. These spectrum ranges have been designated, or "allocated," for satellite services use by nation states participating in ITU world radio communication conferences. Their use is subject to national licensing and reaching coordination agreements with other potentially affected operators in order to ensure operations free from harmful interference from the radio stations of other countries.

Most commercial communications satellites operate in geosynchronous orbits, meaning that they revolve around Earth in the same timespan as Earth's rotation. This is a unique circular orbit almost a tenth of the way out to the Moon made famous by Arthur C. Clarke who first proposed it for communications satellites in 1945 before the Space Age even began. The GSO is the only orbit where satellites appear to hover above the equator as Earth rotates on its axis. This is very helpful because ground antennas do not have to track the satellite but can be constantly pointed to a single satellite. There are many other satellite orbits that are also in use, such as low Earth orbits (below the Van Allen Belts) medium Earth orbits (typically just above the Van Allen Belts), elliptical orbits (that slice through the Van Allen Belts), Su—synchronous orbits, super synchronous orbits, and so on. However, the most popular orbit by far is the GSO.

Overall, 42 % of existing satellites are in a geosynchronous orbit (Satellite Industry Association, State of the Satellite Industry Report 2012, p. 6). The ITU has officially defined the geostationary orbit in terms of its precise physics, although sometimes it is also referred to as the Clarke orbit in honor of Sir Arthur Clarke. A geosynchronous orbit rotates with Earth, but can move up and down north or south from the equator slightly with each rotation. A GSO is a conceptual orbit that does not move off of a sub-equatorial point either east to west or north to south (ITU Radio Regulations, Vol. I, No. 1.189 2012). These geostationary satellites can potentially cover one third of Earth's surface from a height of 35,786 km. The relative geostationary location of the satellite is also beneficial for sending signals to Earth stations with antennas that can be constantly pointed to the satellite without the expense and complexity of tracking systems. As illustrated in Fig. 2.1, the geostationary orbit has areas of intensive occupation, particularly over landmasses where communications services are most developed and most highly sought after by paying customers. However, there is no actual physical limit established for how many satellites can operate in the arc, as it is subject to many factors. In fact, some operators deploy multiple satellites in a single orbital position.

The role of the ITU in managing access to the GSO, together with access to radio frequency spectra, is of central importance to current and future satellite operators and governments and is an area of key focus by the ITU's radio communication sector.

## A Closer Look at the International Telecommunication Union

The ITU is a UN entity, known as a specialized agency. The International Bank for Reconstruction and Development (IBRD or World Bank), the World Health Organization (WHO), the International Civil Aviation Organization (ICAO), the

**Fig. 2.4** The headquarters of the International Telecommunication Union in Geneva, Switzerland, near Palais des Nations (Image courtesy of the ITU)

Food and Agricultural Organization (FAO), the UN Environmental Program (UNEP), and the International Maritime Organization (IMO) are examples of UN specialized agencies.

The ITU is headquartered in Geneva, Switzerland, and has a broad mandate of advancing telecommunications to the world's inhabitants. This role includes facilitating the use of radiofrequency spectra, and their associated orbits, by the nations of the world. For satellite operators, the ITU is the organization that oversees access to the GSO and other orbits through management of the coordination process; maintenance of the Master International Frequency Register (MIFR); development of global standards; and allocation of radio frequency spectra to radio services. Successful completion of the ITU's regulatory processes results in a frequency assignment and associated orbital position for GSO satellites that are recorded with a favorable finding in the MIFR. This status affords the satellite operator with international recognition of its operation consisting of priority over other potential users of these resources and protection from harmful interference from those operators with lesser rights. Such international protections, in addition to national licensing, afford the satellite operator with assurance necessary to support the great investments of money and other resources necessary to construct, launch, and operate a satellite network (Fig. 2.4).

The ITU was established in 1865, originally as the International Telegraph Union, driven by the necessity of European nations to coordinate and standardize the interconnection of new telegraph systems spanning their borders. The advent of radiocommunication; the impetus of promoting safety at sea; and the necessity of managing interference inspired the creation of an International Radiotelegraph Union in 1906, and an agreed set of technical regulations, the "Radio Regulations," under which signatory nations agreed to operate their radio stations in a manner so as to prevent harmful interference to the duly authorized radio stations operating in other countries. The sinking of the *Titanic* in 1912 boosted the interest of nations in regulating the use of radio spectra and in cooperating internationally to improve safety at sea through robust requirements for safety radio communications onboard ships, including protecting designated radio frequencies from harmful interference (Johnson).

Following World War I, and the rapid technical development and implementation of new radio technologies and services (including aeronautical applications and sound broadcasting), the International Radiotelegraph Union continued to meet, formulating many structures and mechanisms echoed in today's ITU, including allocation of radio frequency bands to defined radio services and establishing the right to protection from harmful interference. In 1932, this Union and its separate, sister body, the International Telegraph Union, each convened conferences in Madrid and made the decision to merge their organizations, which was facilitated by the fact that they had similar structures and working methods and each depended on the same Swiss civil bureaucrats to perform their administrative functions. Thus, the International Telecommunication Union was forged, originally based in Berne, Switzerland. The term "telecommunications" was defined broadly at the time to include telephony, telegraphy, radio, and other systems of electrical or visual signaling (Codding and Rutkowski 1982, p. 18).

Upon the conclusion of World War II, the Union convened a series of conferences in Atlantic City in the United States to accommodate the burgeoning new uses of spectra that had developed for broadcasting, radiodetermination, and navigation. The 74 nations that participated in the Atlantic City Telecommunication Conference adopted the ITU's modern-day structure and decided to join the Union to the newly created United Nations as a specialized agency.

The ITU is a periodic international organization, as it seeks to achieve international consensus through decisions of member states entered into at treaty conferences convened on a periodic basis. The governing body of the ITU is the Plenipotentiary Conference, which meets every four years and is empowered to adopt and amend the ITU's basic instruments, the ITU Constitution, and Convention (ITU 2011). The ITU also convenes world radio communications conferences, normally every 3–4 years, to adopt or revise the international Radio Regulations, a treaty which contains technical, operational, regulatory, and procedural provisions governing access to radio frequency spectra and associated orbital resources. Each country has one vote to cast at treaty conferences. However, in most cases, except for elections, decisions are made by consensus, rather than by individual country voting.

In addition to periodic conferences, the Union is composed of several permanent features, shown in Fig. 2.5, including a Secretary-General, who serves as the legal representative of the Union; the Council, a board of directors that meets annually and governs the Union between plenipotentiary conferences; and three substantive sectors—Radiocommunication, Telecommunication Standardization,

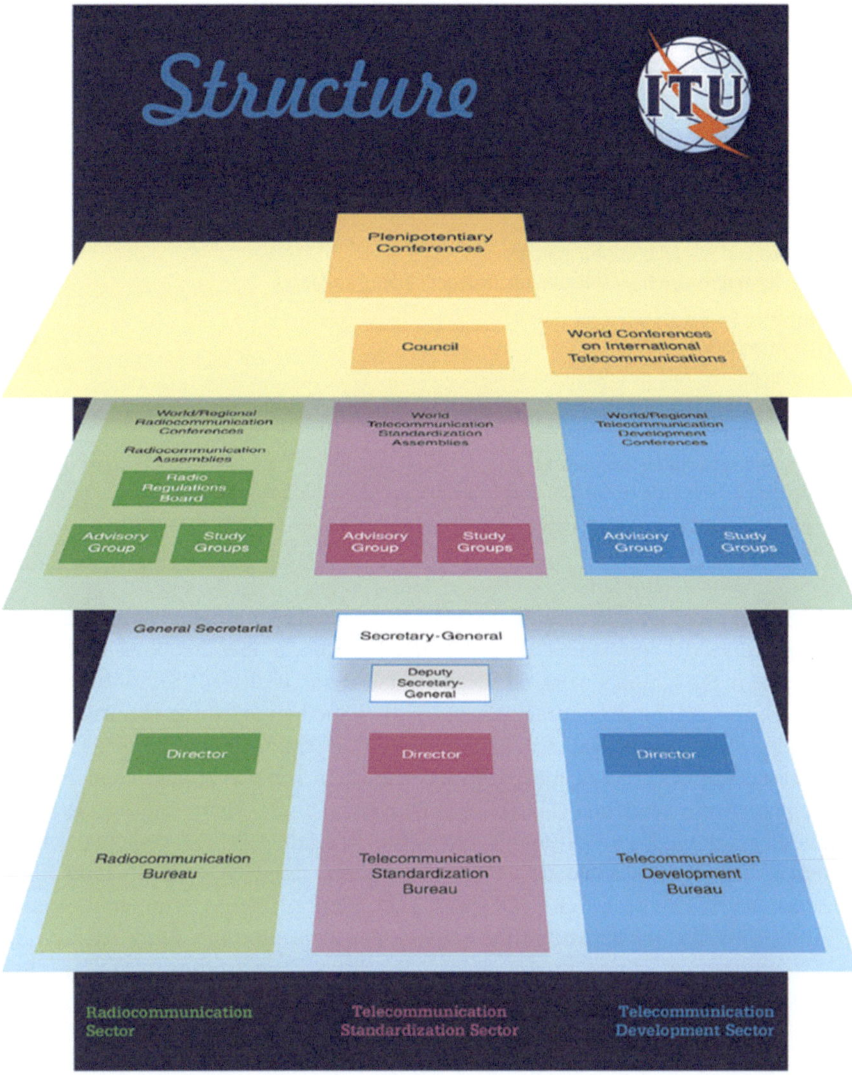


**Fig. 2.5** Structure of the ITU. Featured are three vertical sectors, each lead by a conference or assembly and supported by a director (Image courtesy of the ITU)

and Telecommunication Development—which perform the technical work of the Union, including technical studies, approving standards (called Recommendations), developing handbooks, and making the preparations for treaty conferences.

The membership of the Union is characterized by a blend of government actors and members of the private sector. Only member states carry the right to vote and to actively participate in treaty conferences. State membership in the ITU is nearly universal with 193 country members, the newest member being South Sudan.

**Table 2.2** Minimum annual contributory amounts for other than member states (Image courtesy of the ITU)

| Minimum annual contributory amounts[a] | |
|---|---|
| Standardization/Radiocommunication sectors | |
| Sector members | 31,800 |
| Sector members from developing countries (per capita income not exceeding USD 2,000) | 3,975 |
| Associates | 10,600 |
| Academia and research establishments | 3,975 |
| Academia and research establishments fromdeveloping countries | 1,987.50 |
| Development sector | |
| Sector members | 7,950 |
| Sector members from developing countries | 3,975 |
| Associates | 3,975 |
| Associates from developing countries | 1,987.50 |
| Academia and research establishments | 3,975 |
| Academia and research establishments fromdeveloping countries | 1,987.50 |

[a]All amounts in Swiss francs (CHF)

One of the unique features of the ITU as an intergovernmental organization is that private sector companies are encouraged to join as members and to actively participate in the Union's working-level substantive activities as "Sector Members." These private sector entities may join any or all of the Union's three sectors as sector members, with the approval of their national administration. As of 2013, there were 513 sector members, associates, who participate in a more limited capacity, and participants from academia (ITU Global Directory 2013). Many satellite operators are members of the Radiocommunication Sector, including Telesat, AsiaSat, YahSat, Thuraya, Hispasat, Globalstar, Intelsat, Iridium, Eutelsat, ACeS, INDOSAT, SES, MEASAT, SatMex, NIGCOMSAT, and Telenor. Intergovernmental satellite organizations also participate in the activities of the sector, including ARABSAT, the International Telecommunication Satellite Organization (ITSO), and INTERSPUTNIK.

Sector membership requires a sizable annual financial contribution for each sector, as shown in Table 2.2. Sector membership in the Radiocommunication Sector is currently 31,800 Swiss francs (or CHFs) (approximately $34,000) for companies not from a developed country. The Radiocommunication Sector is responsible for technical standards, managing the satellite coordination process, and maintenance of the Master International Frequency Register. So why would so many commercial satellite operators consider it worthwhile to invest such a large amount of money each year in membership in a UN body in which they cannot even vote? Sector membership entitles them to attend and actively participate in ITU technical activities, including in the Study Groups and their Working Parties, working side by side with the world's regulators, where they can launch studies and introduce contributions leading to formulation of standards and technical reports that establish the basis for treaty conference decisions on spectrum allocations and associated orbital use.

Thus, as the satellite operators develop new technologies and services and seek access to expanded spectra and orbital resources, they can drive the changes to the international regulatory landscape needed to accommodate their business plans.

On the other hand, sector members can also work through ITU processes to define their technical characteristics and protection criteria to ensure that other providers of spectrum-supported services are constrained from operations that could cause harmful interference to their current or planned operations or to limit their future expansion. Thus, sector members have the opportunity to essentially draft the regulations that will be applied to their operations and to the operations of neighboring spectrum users to allow for the maximum extent of operations in their spectrum while also preventing harmful interference from other services. These international standards are often implemented by national regulators in their domestic regulations to serve as a basis for national licensing. Sector members also have the opportunity to learn what their competitors are planning and to impact the plans of others, which could have a deleterious effect on their operations. The decisions of ITU bodies have direct impact on satellite operations.

The ITU's basic instruments guarantee certain rights to sector members to fully participate in the work of the sector, including adoption of questions for study and recommendations and providing chairmen and vice-chairmen for study groups and assemblies (ITU, Collection of Basic Texts, Constitution Article, 3; Resolution 14 (Rev. Antalya 2006)). They may also attend treaty conferences, including World Radiocommunication Conferences, as observers, with more limited participatory rights. However, many nations permit satellite operators to serve on their national delegations to treaty conferences, giving them opportunities to draft proposals and contribute to their delegation positions and negotiations as technical experts. For example, SES participants are often seen speaking on behalf of Luxembourg. This is also a unique arrangement for the United States with the US Congress passing a law to enable corporate participation on its official delegations, although the roles of such corporate participants on US delegations is to provide technical advice.

The ITU, which will celebrate the 150th anniversary of its founding in 2015, is the longest running and most successful intergovernmental organization in history. This success is attributed to the fact that it is a technical, historically non-political body driven by the rapid progress of technology and its importance to society. But another key factor that keeps the ITU relevant is the key role played by its private sector participants who develop and implement so many of the world's technological advances and are driven by competitive forces to grow and to make the most effective and efficient use of available resources.

## The Beginning of International Space Regulation

The ITU began to address space communications in 1959, when it amended the Radio Regulations to add definitions for Earth stations, space stations, and initial space services. It convened its first space treaty conference in 1963, the Extraordinary

Administrative Radio Conference to allocate frequency bands for space radiocommunication purposes ("EARC-63"), in which it began the process of allocating spectrum to new space radio services.

Initially, the space processes were drafted in the same vein as those developed many years earlier for terrestrial services. Administrations would obtain access to spectrum and orbital resources on a per request basis, subject to following the applicable provisions of the Radio Regulations and not causing harmful interference to stations of other countries duly recorded in the Master International Frequency Register ("MIFR"). By the 1960s, developing countries began to realize the importance of space services for their own international and domestic telecommunications needs and to express concerns that the available resources might be exhausted (or prohibitively expensive for them to access) by the time they were ready to utilize them (Allison 1992).

Many developing countries viewed this "first come, first served" process as inherently inequitable, as it favored those who had the resources to claim access in the present, over those who might be unable to claim access until the future, at which point the supply of available resources could be exhausted. As radiofrequency and orbital resources are limited natural resources in outer space and do not belong to any single nation, developing countries believed that it was their right to claim an equal share of these resources, regardless of current need or capability. They wanted to guarantee their future access to these resources through a priori planning. This demand for guaranteed access to the GSO was similar to assertions made in other international fora for equal access to other common resources, such as minerals on the Moon and in deep-sea beds (White 1988; Labrador and Galace 2005).

This call for an equal share to the orbit/spectrum resources culminated in the adoption of the principle of equitable access by the 1973 Plenipotentiary Conference at Málaga-Torremolinos, building on discussions at recent radio conferences and resolutions of the UN General Assembly. Article 33 of the ITU convention (now Article 44 of the ITU constitution) was amended to add a new paragraph providing:

> In using frequency bands for space radio services Members shall bear in mind that radio frequencies and the geostationary satellite orbit are limited natural resources, that they must be used efficiently and economically so that countries or groups of countries may have equitable access to both in conformity with the provisions of the Radio Regulations according to their needs and the technical facilities at their disposal. (International Telecommunication Convention, Málaga-Torremolinos 1973, pp. 20–21)

During the conference's debate on this new provision, the United States proposed to delete the term "equitable access" from this amendment. Although supported by Japan, Indonesia and France, the US proposal was defeated by a vote of 33 to 9, with 21 abstentions (Chairman of Committee Eight 1973).

In 1976, a group of equatorial countries asserted territorial rights to the geostationary orbital positions directly over their landmasses. This claim by the nations of Colombia, Congo, Ecuador, Indonesia, Kenya, Uganda and Zaire (and Brazil as an observer) became known as the Bogotá Declaration. The declaration appeared to conflict in principle with the provisions of the Outer Space Treaty, to which some of these nations were a party, which provides that outer space is not subject to appropriation by claim of national sovereignty. The Bogotá declaration was considered by several ITU conferences and in other UN venues over the years (Lyall and Larsen 2009).

The 1979 World Administrative Radio Conference (Geneva) (WARC-79) adopted a resolution inviting the Council to take all necessary steps to convene a "world space administrative radio conference … not later than 1984 to guarantee in practice for all countries equitable access to the geostationary-satellite orbit and the frequency bands allocated to space services." (Final Acts of the World Administrative Radio Conference, Geneva 1980, p. 744)

At the next plenipotentiary conference in 1982 (Nairobi), the developing country proponents were successful in strengthening Article 33 of the ITU convention to reword the equitable access provision to replace the text "according to their needs and the technical facilities at their disposal" with "taking into account the special needs of the developing countries and the geographical situation of particular countries." (Final Acts of Plenipotentiary Conference, Nairobi 1982, p. 28) The Nairobi conference also validated the WARC-79 agreement to convene a world radio conference to consider planning of spectrum and orbital resources.

The "Space WARC," the World Administrative Radio Conference on the Use of Geostationary-Satellite Orbit and the Planning of Space Services Utilizing It, was held in two sessions in 1985 and 1988 with the mandate to "reconcile the principle of guaranteed and equitable access with that of the efficient and economic use of two limited natural resources: the GSO and the radio frequency spectrum." (ITU 1988, ORB-88 Adopts Plan and Regulatory Provisions for Geostationary Satellites, *Telecommunication Journal*, 55, p. 790) The Space WARCs resolved the matter by establishing a plan to reserve a certain amount of spectrum and predetermined geostationary orbital position for each country in order to guarantee future access to all member states regardless of current need. At the same time, the conference preserved first come, first served processes for the remainder of the resource—the non-planned bands. It also adopted a simplified regime for coordination and management of harmful interference. This grand compromise set the path forward for the ITU's management of space resources that still applies in the next century. In addition, it "reaffirmed the ITU as one of the most successful international organizations" (Smith 1989, p. 400). These two approaches for obtaining access to orbital and spectrum resources—first-come first-served and planning—remain the principal means to obtain access to the geostationary orbit today.

## References

Allison, A.: Meeting the challenges of change: the reform of the International Telecommunication Union. Fed Comm Law J **45**, 491 (1992)

Boeing: Commercial communications satellites in geosynchronous orbit. www.boeing.com/defense-space/space/bss/launch/980031_001.pdf.

Chairman of Committee Eight: Summary Record of the Sixteenth Meeting of Committee 8 (Rights and Obligations), Málaga-Torremolinos. ITU Doc. 432 (1973)

Codding, G., Rutkowski, A.: International Telecommunication Union in a Changing World. Artech House, Norwood, MA (1982)
International Telecommunication Union: International Telecommunication Convention, Málaga-Torremolinos. International Telecommunication Union, Geneva (1973)
International Telecommunication Union: Final Acts of the World Administrative Radiocommunication Conference, Geneva. International Telecommunication Union, Geneva (1980)
International Telecommunication Union: Final Acts of the Plenipotentiary Conference, Nairobi. International Telecommunication Union, Geneva (1982)
International Telecommunication Union: ORB-88 adopts plan and regulatory provisions for geostationary satellites. Telecommunication Journal. 55. International Telecommunication Union, Geneva (1988)
International Telecommunication Union: Recommendation ITU-R V. 431–7. Nomenclature of the Frequency and Wavelength Bands Used in Telecommunications. International Telecommunication Union, Geneva (2000)
International Telecommunication Union: Collection of the Basic Texts of the International Telecommunication Union Adopted by the Plenipotentiary Conference. International Telecommunication Union, Geneva (2011)
International Telecommunication Union: Radio Regulations. International Telecommunication Union, Geneva (2012)
International Telecommunication Union: ITU Global Directory. http://www.itu.int/online/mm/scripts/mm.list?_search=SEC&_languageid=1 (2013)
Johnson, D.: The Radio Legacy of the R.M.S. Titanic. http://www.avsia.com/djohnson/titanic.html
Labrador, V., Galace, P.: A brief history of the satellite communications industry. Sat Magazine (2005)
Lyall, F., Larsen, P.: Space Law: A Treatise. Ashgate, Surrey (2009)
Satellite Industry Association: State of the Satellite Industry Report. http://www.sia.org/wp-content/uploads/2012/06/Final-2012-State-of-Satellite-Industry-Report.pdf (2012)
Satellite Industry Association: State of the satellite industry report. http://www.sia.org/wp-content/uploads/2013/10/2013_SSIR_Final_Oct.pdf (2013)
Smith, M.: International regulation of satellite telecommunications after the space WARC. Dissertation (D.C.L.), McGill University (1989)
White, L.: Law of International Space Communications. Artech House, Norwood, MA (1988)

## *Websites*

http://transition.fcc.gov/omd/history/radio/quality.html
http://www.sia.org/wp-content/uploads/2011/10/SATELLITE_101_2011.pdf

# Chapter 3
# International Regulatory Framework for Satellites

## ITU Regulatory Framework for Satellites

Access to orbital and associated spectrum resources is managed through two separate processes contained in the ITU'S Radio Regulations. These processes are established by nation states at treaty conferences, typically at world radiocommunication conferences. The two processes are generally referred to as "first come, first served," which is detailed in Articles 9 and 11, in Volume I of the Radio Regulations; and planning (or a priori), contained in Appendices 30, 30A and 30B in Volume II of the Radio Regulations. The original process, first come, first served is based on coordination among nations and registration of assignments in the MIFR, which status conveys international recognition and protection from harmful interference from other nation's stations. The plans contain reservation of orbital resource and bandwidth for each country, together with operational parameter, thus serving to guarantee future access to all nations. These processes may seem complex and difficult, and that is because they truly are (Fig. 3.1).

Radiocommunications engineers and attorneys with specialized training have to spend years to become expert on these procedures. This is further complicated by the fact that the ITU has divided the world into three parts. These are Region One (Europe, the Middle East and Africa); Region Two (the Americas) and Region Three (Asia and Australasia). Region Two, the Americas, is the least planned of the regions (Fig. 3.2).

The planning process grew out of the concerns of developing countries for guaranteed access to orbital/spectrum resources into the future for provision of national satellite services. Adoption of the plans was extremely challenging, both in political and technical respects. The plans were developed and implemented over a period of decades and successive conferences. The first space plan was for the Broadcasting-Satellite Service (BSS). The BSS plan for Regions 1 and 3 was adopted in 1977 by the World Administrative Radio Conference for the Planning of the Broadcasting-Satellite Service in the Frequency Bands 11.7–12.2 GHz (Regions 2 and 3) and 11.7–12.5 GHz (Region 1) (Geneva). The Region 2 countries did not agree to this

A.L. Allison, *The ITU and Managing Satellite Orbital and Spectrum Resources in the 21st Century*, SpringerBriefs in Space Development,
DOI 10.1007/978-3-319-05314-1_3, © Audrey Allison 2014

**Fig. 3.1** The ITU gathers in 2006 to celebrate the centenary of the Radio Regulations (ITU File Photo)

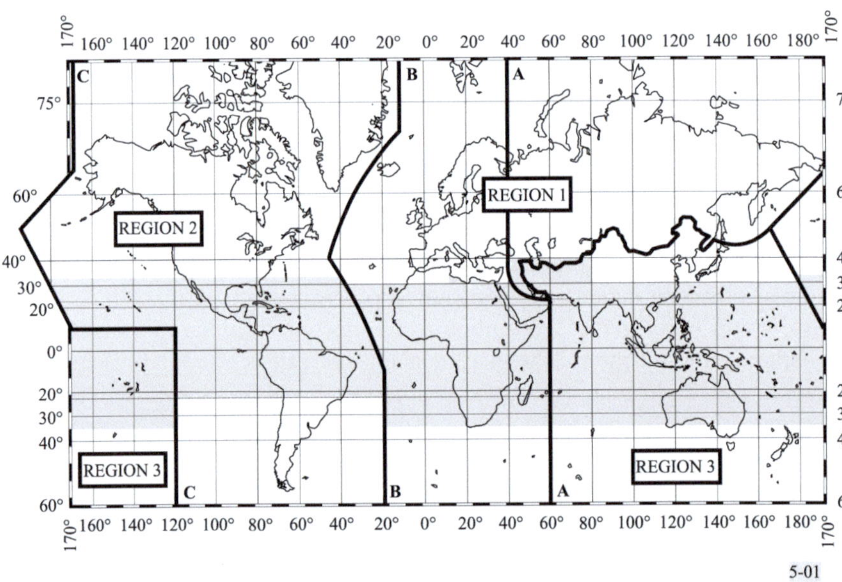


**Fig. 3.2** Demarcations of radio regions for purposes of frequency allocation in the Radio Regulations (ITU Radio Regulations, No. 5.2, Geneva 2012)

plan and later adopted a separate, more flexible plan at the 1983 Regional Administrative Conference for the Planning of the Broadcasting-Satellite Service in Region 2 (Geneva). These plans cover the downlinks for the BSS. Separate feeder links plans were adopted in 1985 following the allocation of spectrum for this purpose. A single worldwide plan for fixed-satellite service (FSS) was adopted in the new C and Ku-band allocations by the 1988 World Administrative Radio Conference on the Use of Geostationary-Satellite Orbit and the Planning of the Space Services Utilizing It (Geneva, WARC ORB-88). The plans were later updated by World

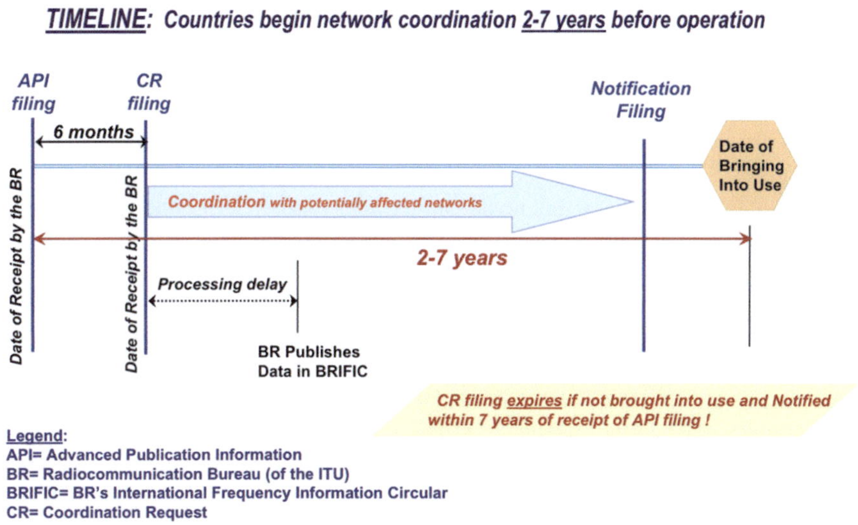


**Fig. 3.3** The ITU's satellite coordination process timeline (Image courtesy of Boeing)

Radio Conferences in 1997, 2000 and 2007 to include provision for additional countries, additional capacity, and other changes to improve usability.

Although the purpose of the WARC ORB-88 was for space planning, the conference was ultimately able to achieve consensus by reaching a compromise between the interests of developing countries in guaranteed access to space and the interests of the developed world and current satellite system operators in retaining the current demand-based process. Thus, the first come, first served regime for geostationary satellites in the FSS was retained for the non-planned frequency bands. Under the first come, first served process, the basis of ensuring interference-free operation of satellite systems is built upon nations securing rights through adherence to the Radio Regulations and following its detailed procedures for coordination and registration in the MIFR. This process involves indentifying which assignments with greater rights may be affected by a proposed satellite network; various approaches for calculating potential interference; and a standard series of procedures for communicating and responding to information among members states and the radiocommunications bureau (Smith 1989).

The detailed steps for coordinating a satellite network differ for various services, frequency bands, and orbits. A satellite network is defined as a satellite and a cooperating Earth station. For the typical geostationary satellite case, the basic common steps for obtaining access to the orbit in an unplanned band are described in Fig. 3.3.

The process begins with the electronic submission of Advance Publication Information (API) to the ITU's Radiocommunication Bureau by notifying the administration (the country representing the operator) with a general description of the proposed network (containing the elements described in Appendix 4 of the Radio Regulations). The API is supposed to be filed no earlier than 7 years and preferably no later than 2 years before the planned date of bringing into use the satellite network (ITU Radio Regulations 2012, No. 9.1). The information is electronically published in the bureau's International Frequency Information Circular (BR IFIC). The date of the bureau's receipt of the API "starts the clock" of the 7-year time period provided by the regulations for completing coordination and for bringing the satellite into service, known in ITU parlance as "bringing into use." If the administration fails to complete the project within the 7 years, it is subject to cancellation by the bureau—meaning that it will no longer retain its place in the queue and no longer be protected from harmful interference from later-filed networks.

Within 2 years of the bureau's receipt of the API, the notifying administration must submit its Coordination Request (CR) to the bureau. The CR is required to include complete technical data for the network as called for in Appendix 4 of the Radio Regulations. Failure to submit the CR within this period also results in the cancellation of the filing. The date of the bureau's receipt of the CR establishes the network's place in the "queue" of filings and its priority over later filings. A proposed new network must be capable of avoiding harmful interference to those networks already recorded in the MIFR and also to those planned networks residing earlier in the queue and later brought into use and recorded in the MFIR. The technical considerations supporting coordination are described in Article 9 and Appendix 5 of the Radio Regulations, and are supported by innumerable ITU-R recommendations, reports and handbooks.

The Radiocommunication Bureau electronically publishes the coordination data in an information circular after which point the network proponent carries out the required coordination with potentially affected networks during the remainder of the 7-year period. The bureau is available to assist administrations with this process, which includes several steps of correspondence between parties.

Pursuant to the provisions of Article 11 of the Radio Regulations, the notifying administration must submit its notification to the bureau that the required coordination has been successfully completed and that the network has been brought into use by the conclusion of the 7-year deadline, else the filing will be canceled by the bureau. In addition, as discussed further in Chaps. 6 and 7, the notifying administration must submit Resolution 49 due diligence data and it (or the operator) must pay satellite network cost recovery fees to the Bureau. Typical fees for a geostationary communications satellite are: CHF 570 for API; CHF 24,620 for a CR; and CHF 30,910 for Notification (ITU Cost Recovery Invoices for Satellite Network Filings 2012).

Upon receipt of the notification, the bureau will examine the filing for completion of coordination and conformity with the Radio Regulations. If the results of this examination are favorable, the network is recorded in the MIFR. It is the recording of the frequency assignment in the MIFR that confers legal status to the network so that it receives international recognition of its frequency assignment and the right to be protected from harmful interference from other radio stations. This is the culmination of the ITU's coordination process.

It has been casually observed that "obtaining access to the most vital frequencies [and orbital slots] is seen as high art by some and as voodoo by others" (C. Stott, SatMagazine 2009). However, over the past 50 years, the industry has been able to grow and thrive, in part due to the stability provided by this imperfect, yet evolving international regulatory framework. The next chapter discusses how the process ceased to be effective in the latter part of the twentieth century, and how the world came together to improve it.

## Equitable Access

Article 44 of the ITU's constitution provides the bedrock principle of the use of the orbits and spectrum resources for satellites. The language of Article 44 provides:

> Use of the Radio-Frequency Spectrum and
> of the Geostationary-Satellite and
> Other Satellite Orbits
>
> Member states shall endeavour to limit the number of frequencies and the spectrum used to the minimum essential to provide in a satisfactory manner the necessary services. To that end, they shall endeavour to apply the latest technical advances as soon as possible.
>
> In using frequency bands for radio services, Member States shall bear in mind that radio frequencies and any associated orbits, including the geostationary-satellite orbit, are limited natural resources and that they must be used rationally, efficiently and economically, in conformity with the provisions of the Radio Regulations, so that countries or groups of countries may have equitable access to those orbits and frequencies, taking into account the special needs of the developing countries and the geographical situation of particular countries. (ITU Collection of Basic Texts 2010, p. 42)

This provision reflects the union's dual streams of the interest of: (1) promoting the most efficient use of the common shared resources of space and spectrum through employment of the latest technical advances, and (2) ensuring that countries have equitable access to those resources. What does "equitable access" mean exactly? That has been a matter of considerable debate in the ITU since the introduction of space services. It is a matter that is reconsidered at every plenipotentiary and World Radiocommunication Conference.

Consideration of equitable access should be undertaken in the context of the 1967 Treaty on Principles Governing the Activities of Space in the Exploration and Use of Outer Space, including the Moon and other celestial bodies ("the Outer Space Treaty"), which establishes that outer space, including the orbits, is the province of all humankind and not subject to national appropriation by means of use or occupancy. The ITU's treaties on use of the geostationary and other orbits define a legal regime within the framework of the Outer Space Treaty to establish the rights of states to access space in a rationale, stable, and predictable regulatory environment necessary to support the substantial investment of resources required for the manufacture, launch and operation of a satellite network (Smith 1989). These investments are necessary to support development of services for the benefit of the world's inhabitants. The UN Committee on Peaceful Uses of Outer Space (COPUOS) has recognized the ITU and its treaty provisions for ensuring equitable access to geostationary orbit (United Nations 2000).

The essence of these provisions is that access to frequencies and orbits in space cannot be "bought" or "sold." Countries can only derive the right to use these resources through the defined processes of the ITU. In later chapters we will see how efforts were made to circumvent these international procedures, in a process known as "paper satellites." Chapter 4 addresses the nature of these efforts to go around the rules and what corrective actions have subsequently been taken.

The Space WARCs sought to satisfy the concerns of developing countries for equitable access by guaranteeing access to spectrum and orbital resources through adoption of plans for Broadcasting-Satellite and FSS. These plans are contained in the appendices to the Radio Regulations (Appendices 30, 30A and 30B). However, as the years pass, these plans, which were premised on the state of technology at the time of planning, have become outmoded, inefficient, and difficult to implement and to modify. At this time, the vast majority of planned resources remain fallow. In the decades following the Space WARC, developing countries have continued to express concerns over their equitable access to the orbit, with repeated complaints aimed at the procedures used for access to the unplanned bands.

Twenty-one years after the Bogotá Declaration, Colombia, joined by Ecuador and Costa Rica, submitted a proposal to the 1997 World Radiocommunication Conference in Geneva under the agenda item relating to consideration of Resolution 18 (Kyoto 1994), *Review of the ITU's frequency Coordination and Planning Framework for Satellite Networks* (see Appendix A and Chap. 4, below). The three countries proposed that, as part of the Resolution 18 review, it would be necessary to take into account the principles of Article 44 of the ITU constitution and to establish "due diligence" measures for their application to the coordination and notification of satellite networks (ITU Doc. CRM97/203, 1997, p. 1). Their proposal included a draft new resolution on *Due diligence in applying the principles embodied in the Constitution*, which noted that the preamble to the Radio Regulations recounts that a founding principle of the Radio Regulations is:

> In using frequency bands for radio services, Members shall bear in mind that radio frequencies and the geostationary-satellite orbit are limited natural resources and that they must be used rationally, efficiently and economically, in conformity with the provisions of these Regulations, so that countries or groups of countries may have equitable access to both, taking into account the special needs of the developing countries and the geographical situation of particular countries. (No. 196 of the Constitution) (ITU Radio Regulations, Vol. I, p. 3)

Colombia, and its two Latin American colleagues, proposed that the ITU apply these principles in applying Section II of Article 11 of the Radio Regulations on examination of notices and recording of frequency assignments in the Master Register. The proposal sought to instruct the ITU's Radio Regulations Board (RRB), a panel of elected subject matter experts who assist the bureau, to develop Rules of Procedure to implement this concept. Furthermore, it proposed that the bureau immediately cease recording frequency assignments until such new procedures could be applied. Finally, it provided that the new procedure, once adopted, should be applied to all notices, even those previously recorded, to verify compliance with the due diligence procedure (ITU Doc. CMR97/203 1997, p. 2).

The conference gave due consideration to this potentially destabilizing proposal. The United States, which shared many common inter-American proposals with the proponents, collaborated with Colombia and Ecuador to reach a constructive solution. The conference ultimately approved a streamlined version of the proposal, henceforth known as Resolution 80, calling for development of rules of procedure on the notification provisions in light of the preamble of the Radio Regulations for further consideration by the next World Radiocommunication Conference (Final Acts of the World Radiocommunication Conference, Geneva 1997).

The next conference took place in 2000 in Istanbul. WRC-2000 considered Resolution 80 and a report developed by the RRB in response to the resolution. The only progress that the conference was able to achieve in this delicate matter was to add a summary of the RRB's report into a slightly expanded version of the resolution, stating that:

- The "first-come first-served" concept restricts and sometimes prevents access and use of certain frequency bands and orbit positions.
- A relative disadvantage for developing countries in coordination negotiations due to various reasons such as a lack of resources and expertise.
- Perceived differences in consistency of application of the radio regulations.
- The submitting of "paper" satellites that restrict access options.
- The growing use of the bands of the plans of Appendices [S]30 and [S]30A by regional, multichannel systems, which may modify the main purpose of these plans to provide equitable access to all countries.
- The considerable processing delays in the Radiocommunication Bureau, due to the very complex procedures required and the large number of filings submitted. These delays contribute to a coordination backlog of 18 months, which could extend to three years and creates uncertain regulatory situations, additional delay in the coordination process that cannot be overcome by administrations, and the possible loss of the assignment because the allotted time is exceeded.
- Satellite systems that may already be in orbit before completion of coordination.
- Statutory time-frames, such as in [S]11.48, that may often be insufficient for developing countries to be able to complete the regulatory requirements as well as the design, construction and launch of satellite systems.
- A lack of provisions for international monitoring to confirm the bringing into use of satellite networks (assignments and orbits). (Resolution 80, Rev, 1997)

WRC-2000 resolved to instruct the ITU's Radiocommunication Advisory Group, an advisory body to the director, to carry out further studies on the linkage of the notification, coordination, and recording procedures with the principles of Article 44 (equitable access). In addition, the RRB and the director of the Radiocommunication Bureau were charged to make additional reports to the following World Radiocommunication Conference on progress made on these taskings. And thus this pattern continued from WRC to WRC (ITU, Final Acts of the World Radiocommunication Conference, Geneva 2000).

Following the next World Radiocommunication Conference in 2003 (WRC-03 2003) in Geneva and the continued absence of substantive progress, Resolution 80 was again revised by the 2007 World Radiocommunication Conference to add complete sections of the RRB's report to WRC-2000 and its report to WRC-03 in annexes to the resolution. The 2007 revision to Resolution 80 instructs the radiocommunication sector to carry out studies on procedures for measurement and analysis of the application of the basic principles contained in Article 44 and the director to make detailed progress reports to each future World Radiocommunication Conference (Final Acts of the World Radiocommunication Conference, Geneva 2007).

Although the work of Resolution 80 remains incomplete to this day, it remains, like "equitable access," a work in progress (ITU-R: Contribution to the report on the long-term sustainability of outer space activities 2012). Consideration of Resolution 80 is now a standing agenda item at every WRC, thus ensuring the consideration of equitable access anytime the world gathers to review and update the procedures for access to the spectrum and orbital resources.

## References

Colombia: Proposals for the Work of the Conference, Geneva. ITU Doc. CMR97/203-E (1997)

International Telecommunication Union: Final Acts of the World Radiocommunication Conference, Geneva. International Telecommunication Union, Geneva (1997)

International Telecommunication Union: Final Acts of the 2000 World Radiocommunication Conference, Istanbul. International Telecommunication Union, Geneva (2000)

International Telecommunication Union: Final Acts of the 2003 World Radiocommunication Conference, Geneva. International Telecommunication Union, Geneva (2003)

International Telecommunication Union: Final Acts of the 2007 World Radiocommunication Conference, Geneva. International Telecommunication Union, Geneva (2007)

International Telecommunication Union: Collection of the Basic Texts of the International Telecommunication Union Adopted by the Plenipotentiary Conference. International Telecommunication Union, Geneva (2011)

International Telecommunication Union: Radio Regulations. International Telecommunication Union, Geneva (2012)

International Telecommunication Union: Cost Recovery Invoices for Satellite Network Filings. http://www.itu.int/ITU-R/go/space-cost-recovery/en (2012)

ITU-R: Contribution to the report on the long-term sustainability of outer space activities. http://www.itu.int/en/ITU-R/space/snl/Documents/ITU-Space_reg.pdf. (2012)

Smith, M.: International regulation of satellite telecommunications after the space WARC. Dissertation (D.C.L.), McGill University (1989)

Stott, C.: In my view: thoughts on satellite filings. SatMagazine. http://www.satmagazine.com/storyphp?number=1369991233 (2009). Accessed November 2009

United Nations: Report of the Committee on Peaceful Uses of Outer Space. United Nations, New York (2000)

## *Websites*

ITU Space Services Department: http://www.itu.int/ITU-R/go/space/en

# Chapter 4
# The Advent of Paper Satellites

## The Backlog

In the mid-1990s, as commercial satellite projects were proliferating and as successive World Radiocommunication Conferences (WRCs) were making new global spectrum allocations and other regulatory provisions to accommodate these new satellite services, a veritable gold rush of satellite network filings ensued. The ITU's Radiocommunication Bureau was accumulating vast piles of satellite network filing paperwork to process and, consequently, a growing backlog of networks to process and publish (Fig. 4.1).

As observed by the director of the Radiocommunication Bureau in his report to the 1997 WRC (WRC-97):

> For some time now, the demand for spectrum/orbit usage has been increasing dramatically, practically for all space communication services and in particular for fixed-satellite, broadcasting satellite and mobile-satellite (GSO or non-GSO) services. In some regions, in the most popular satellite frequency bands, it is becoming increasingly difficult to coordinate satellite networks. Coordination negotiations are becoming long and difficult, and in some cases, satellites are being brought into use or re-positioned without proper coordination. There are some regions, with rapidly developing telecommunication sectors and particularly "valuable" orbital segments, where real conflicting situations are emerging due to the limited capacity of the spectrum/orbit resource. (ITU Doc. CMR97/8-E 1997, p. 2)

Structural changes in the satellite industry were one contributor to the increase in ITU filings (ITU Doc. CMR97/8-E 1997, p. 6). As the role of traditional public monopoly operators declined, and the industry became characterized by privatized and new private commercial satellite systems, access to resources becoming more a matter of commercial competition and less a matter of good will between nation states. Also contributing to the backlog was the growing technical complexity of proposed satellite systems and the dawning recognition by governments and the commercial sector of the economic importance of satellite communications. But perhaps the most significant cause of the backlog was "Paper Satellites."

A.L. Allison, *The ITU and Managing Satellite Orbital and Spectrum Resources in the 21st Century*, SpringerBriefs in Space Development,
DOI 10.1007/978-3-319-05314-1_4, 

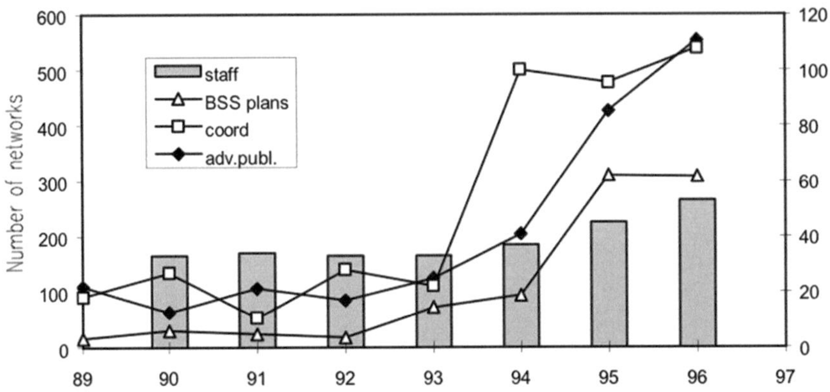


**Fig. 4.1** Evolution of space submissions (by advance publication, coordination request, and BSS Plan filings) and ITU staff in the 1990s (ITU Doc. CMR97/8-E 1997, p. 5)

## Paper Satellites

A major contributor to the growing backlog of satellite network filings was the increasing practice of "overfiling," also politely known as "the reservation of capacity without actual use." Eventually the common term for such overfilling became "paper satellites," because these satellites were only regulatory filings on reams of paper sent to Geneva and were not intended for actual manufacture and launch by the filing party.

This paper satellite period was marked by the filing of coordination requests for more orbital positions and frequency assignments than was actually planned to be implemented. Among the reasons for paper filings was maximizing flexibility, warehousing resources, forestalling competition, and/or simply trying to "cash in." As the backlog developed and uncertainty grew, operators were further motivated to "overfile" as a means of hedging their bets that one filing would be enough as they attempted to overcome the uncertainty of success of any particular filing. Thus the problem snowballed.

In one particularly infamous instance of paper satellite filing, the small Pacific island state of Tonga allowed an individual, an ex-official of Intelsat familiar with ITU filing processes, to form a company called "Tongasat," which filed a series of speculative satellite coordination filings with the ITU for prime orbital locations. The purpose of these multiple filings was not to construct the satellite networks they described but for the purpose of "selling" coordinated access to space for a profit (reportedly, US$2 million apiece) (Andrews 1990). This speculative scheme launched by Tongasat in 1988 filed for 16 slots crucial to connecting Asia and the United States, and it became emblematic of the gold rush in GSO filings. In 1993, Tonga sought to make a series of aggressive changes to the radio regulations to speed up the coordination process and to remove obstacles to coordination and registration of new satellite networks.

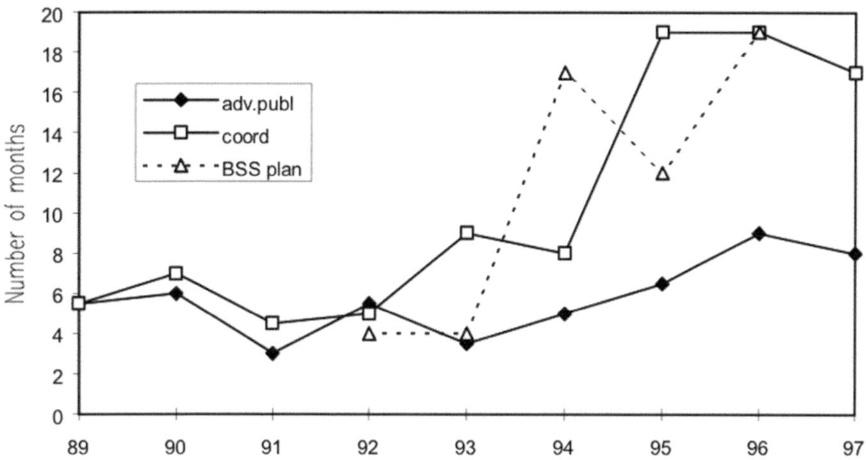


**Fig. 4.2** Delays in ITU publication in the 1990s (ITU Doc. CMR97/8-E 1997, p. 5)

Meanwhile, as Tongasat was making a new business of satellite slot speculation, Australia was unable to complete coordination of its satellite networks in one segment of the geostationary orbit arc. At the 1994 Plenipotentiary Conference, Australia proposed that these difficulties of coordinating new networks in the GSO should be considered by a group of experts. Australia's efforts resulted in a Plenipotentiary Conference action to launch a major review the ITU's satellite coordination and planning framework. This activity was described in a conference action contained in Resolution 18 (Australia 1994; ITU 1994).

Whatever the cause of the congestion and backlog, the result was that it became increasingly difficult for real planned satellite networks to complete coordination within the regulatory timeframe, then 9 years, and within the timelines of the filers' own business plans. Additionally, the validity of the data within the ITU's databases was becoming increasingly unreliable and not representative of actual assets requiring protection from harmful interference. A concern also arose among future service providers, including developing countries. They saw that future opportunities for access to prime orbital slots in the GSO were being diminished or precluded by these fictional paper satellite filings, thus depriving them of equitable access to these resources in the unplanned bands in the future.

The tsunami of satellite network filings also overwhelmed the ability of the ITU's Radiocommunication Bureau to process them and to make the required communications and publications within the time periods required under the radio regulations. In his report to WRC-97 pursuant to Resolution 18, the director of the Radiocommunication Bureau reported 2,200 cases of satellite network processing before the Bureau (ITU Doc. CMR97/8-E 1997, p. 4). The figures from the report, shown below, illustrate the trending increases in the number of satellite network filings and the attendant delays in publication of requests (Fig. 4.2).

The increased number of filings and almost year-long delay in publishing coordination requests had negative consequences to satellite operators and introduced great uncertainty into the process. The number of systems also served to greatly increase the number of parties with whom to coordinate. The challenge was to determine which systems were real, and truly needed to be coordinated with and protected from harmful interference, versus which were just "paper" and could perhaps be ignored. These delays negatively impacted the ability to complete coordination within the regulatory period of bringing the satellite into use (Virlini 2010).

## Seeking a Solution

By the mid-1990s it was widely recognized that the ITU's regulatory system for managing access to the geostationary satellite arc was broken. Prospective satellite network operators were unable to file, coordinate, and launch new satellites with confidence that they would have priority over other filings, and thus protection from harmful interference, in the absence of knowing which filings were ahead of them in the queue awaiting processing in Geneva. At the time, it was thought that either the process had to be substantially reformed, or an entirely new mechanism needed to be created. In discussions beginning in 1994 at the Kyoto Plenipotentiary Conference and continuing through the 2002 Plenipotentiary Conference in Marrakesh, the ITU, its member states and sector members developed a series of regulatory solutions to address these problems, as no better alternative to the ITU for managing access to the orbit and spectrum resources emerged (ITU 1994).

As we will explore in the following chapters, the solutions developed by the ITU community were neither revolutionary nor a radical departure from the ITU's traditional approaches. Instead, as is so often the case in the ITU's history, the solutions arrived upon were a series of incremental changes, developed through negotiation, study and gradual consensus building over a period of successive conferences. Although the subjects and proposed solutions were highly controversial at the time, no votes were cast to arrive at these final decisions. And when some solutions didn't work upon application, they were changed and improved.

The solutions consist of two separate new regimes, one administrative and one financial, which, combined with a number of smaller reforms, ameliorated the problems of overfilling, paper satellites, and backlog. The union was able to reach these results without abandoning or making wholesale changes to the first-come, first-served regime for satellite network filings in the unplanned frequency bands. These reforms even contributed positive outcomes with respect to the union's financial health and promoted furtherance of equitable access. This book separately traces the development of each approach.

Once the Plenipotentiary Conference empowered the WRC with clear legal authority to act, the administrative solution was adopted and implemented by a series of WRCs adding "administrative due diligence" requirements to the satellite regulatory process. The WRCs also enacted a series of regulatory improvements to tighten coordination requirements and to introduce additional transparency in the implementation of satellite networks.

While the ITU was addressing problems related to paper satellites, it was simultaneously facing challenges to its long-term financial viability. Unrelated to the paper satellite issue, the ITU adopted a cost recovery scheme of charging fees to recover the costs of some of the services it provides. Although the ITU expressly rejected using financial approaches for redressing the overfilling problem, it did decide to try to recover its costs for processing satellite network filings as a means of firming up the financial base of the union and spreading costs to a wider body of beneficiaries of ITU services (i.e., the commercial sector). Ultimately, the "satellite network cost recovery" program was also credited with facilitating the union's success in easing the backlog and the paper satellite crisis.

The ITU succeeded in curing the processing backlog and curbing the overfilling problem of the 1990s without harm to a growing industry and the needs of sovereign states. However, at this time, we are again finding the congestion of the system and the number of outstanding filings to be a major challenge for satellite operators, and for administrations planning to launch new networks. In fact, the recent trend is that most satellite networks are being notified without having completed required coordination with potentially affected administrations.(ITU Doc. CMR12/4 [Add.7] [Add.1]-E 2012, p. 3) "Virtual satellites" remain an endemic problem for satellite operators—"virtual" rather than "paper" because ITU filings and publications are now made through electronic means, rather than via paper. As with paper satellites, the union and its members still struggle to find workable and agreeable solutions to effectively manage access to the geostationary orbit. Although some now call for an overhaul of the entire regulatory framework in light of these mounting challenges, others are satisfied with the current stable regulatory regime and remain confident that small changes are all that is necessary to address these difficulties at each WRC.

## References

Andrews, E.: Tiny Tonga seeks Satellite Empire in space. New York Times. http://www.nytimes.com/1990/08/28/business/tiny-tonga-seeks-satellite-empire-in-space.html?pagewanted=all&src=pm (1990). Accessed 28 August

Australia: Proposals for the Work of the Conferences, Kyoto. ITU Doc. PP-94/42-E (1994).

Director, Radiocommunication Bureau: Analysis of the space frequency assignments recorded in the Master International Frequency Register (MIFR), Geneva. ITU Doc. CMR12/4 (Addendum 7) (Addendum 1)-E (2012)

Director, Radiocommunication Bureau: Report on Resolution 18 (Kyoto, 1994), Geneva. ITU Doc. CMR97/8-E (1997)

International Telecommunication Union: Final Acts of the Plenipotentiary Conference, Kyoto. International Telecommunication Union, Geneva (1994)

Virlini, G.: Paper satellites: A puzzle for the industry. Space News. http://www.satellitetoday.com/telecom/2010/01/01/paper-satellites-a-puzzle-for-the-industry/ (2010). Accessed 1 Jan 2012

# Chapter 5
# Due Diligence Solutions

## Australia Pushes for Reform: 1994

In the 1990s, Australia was frustrated with the difficulties it was experiencing in seeking to coordinate satellites in a particular portion of the geostationary orbit arc over the Pacific. Australia noted that the international system of coordination and notification of satellite services had historically been based on satellite services being offered by governments as public services. There were separate national or regional systems for national coverage, and international services were provided jointly by governments cooperating through treaty-based organizations. Australia observed:

> The present reality is much more diverse and dynamic, with the globalization and commercialization of communication systems, including alliances and mergers between national and commercial operators, and the development of entirely new satellite systems. (ITU Doc. PP-94/42-E 1994, p. 5)

In 1994, the ITU convened a Plenipotentiary Conference, the highest deliberative and law-making body of the union, in Kyoto, Japan. At the behest of Australia, that conference considered the growing challenges to the ITU's satellite network filing process in an environment of increasing globalization and privatization of telecommunications networks, particularly satellite networks. Upon presenting its proposal at the Kyoto Conference, Australia stressed that:

> … . [T]he basic elements of the ITU's frequency coordination and planning framework for satellite services had been in existence for almost a quarter of a century and that it was high time to review them in order to align them with modern conditions and ensure that the tensions and obstacles already being felt in that sphere in the Asia and Pacific Region did not spread around the world. (ITU Doc. PP-94/231-E 1994, p. 6) (Fig. 5.1)

Germany agreed with Australia that such a review was a good idea, but asserted that there was a need to be pragmatic and to look for the simplest solution. Such study should be performed through existing ITU mechanisms for consideration and expertise by a WRC. This view was supported by the Netherlands, France, Russia, Korea, Switzerland and the United States. The US delegation was seriously concerned

A.L. Allison, *The ITU and Managing Satellite Orbital and Spectrum Resources in the 21st Century*, SpringerBriefs in Space Development, DOI 10.1007/978-3-319-05314-1_5, © Audrey Allison 2014 

**Fig. 5.1** The ITU's 1994 Plenipotentiary Conference is convened in Kyoto, Japan (ITU File Photo)

about the Australian proposal, but was unable to dissuade others from supporting a study. India, Morocco and Nigeria supported Australia's proposal and urged other UN bodies, including the Committee on Peaceful Uses of Outer Space, to be brought into the study.

In Resolution 18 (Kyoto 1994), *Review of the ITU's Frequency Coordination and Planning Framework for Satellite Networks* (attached in Appendix A), the Kyoto conference addressed the growing concern about accommodation of new satellite networks and the need to maintain the integrity of the ITU's procedures. The conference instructed the director, Radiocommunication Bureau, in conjunction with the various elements of the radiocommunication sector, to initiate a review in time for the 1997 World Radiocommunication Conference (WRC-97), including:

- Linkages between ITU procedures and commitments to take up notified frequencies and orbital positions.
- The ongoing need for the ITU's frequency coordination and planning framework for satellite networks to continue to be relevant to rapidly advancing technological possibilities.

With the objectives of:

- Ensuring equitable access to the radio-frequency spectrum and the geostationary-satellite orbit, and the efficient establishment and development of satellite networks.

- Ensuring that international coordination procedures meet the needs of all administrations in establishing their satellite networks, while at the same time safeguarding the interests of other radio services.
- Examining technological advances in relation to the allotment plans with the aim of determining whether they foster flexible and efficient use of the radio-frequency spectrum and geostationary-satellite orbit (Final Acts of the 1994 Plenipotentiary Conference 1994, pp. 155–56).

## The Due Diligence Debate

In the year following Kyoto, the 1995 World Radiocommunication Conference addressed the instructions of the plenipotentiary and established a work program of studies on the topics outlined in Resolution 18, including studies by the ITU-R study group on satellite services, Study Group 4; the Special Committee on Regulatory/Procedural Matters; the Radiocommunication Advisory Group; and the Radio Regulations Board (RRB), to be completed in time for the WRC-97 (ITU Final Acts of the World Radiocommunication Conference, Geneva 1995). The outcome of these efforts was later gathered into the Radiocommunication Bureau director's report to WRC-97, which took place 1 year before the next scheduled plenipotentiary conference.

The various preparatory efforts for WRC-97 included consideration of a broad range of possible solutions to the problems of orbital congestion and overfiling, but were unable to reach consensus on any particular solutions in advance of the conference. The main approach under consideration was to impose a "due diligence" requirement on satellite network filings to winnow out those filings that were fictional. Under such an approach, satellite network filings would be required to be supplemented either with supporting information to confirm or corroborate the seriousness of the proposal (in other words, to support that it represented a "real" rather than a "paper" satellite network), or with the payment of a fee or deposit that would also serve to indicate the seriousness of the proposer's intentions. This due diligence process could thus employ administrative (also known as procedural) mechanisms or financial mechanisms (ITU Doc. CMR97/8-E 1997). A wide range of additional regulatory measures was also considered.

In the extensive conference preparatory efforts leading up to WRC-97, participants were able to forge an agreed general approach to "administrative due diligence" under Resolution 18, by which a satellite operator, through its notifying administration, would submit supporting information about the details behind its planned network, such as the identity of the spacecraft manufacturer, the contracted date of delivery, the launch vehicle provider, and date of launch, in order to substantiate its proposal. It was left to the conference to agree upon the details of this approach in addition to its implementation, including treatment of the consequences of failure to submit such due diligence information on a timely basis.

With respect to "financial due diligence" approaches, the director of the Radiocommunication Bureau described in his report to WRC-97 that there were three potential financial approaches that were also developed during the preparatory cycle:

1. Requiring notifying administrations to pay a filing fee to recover the ITU secretariat's costs for processing satellite network filings.
2. Requiring payment of annual registration fees to maintain those satellite network registrations in the Master International Frequency Register (MIFR).
3. Requiring a deposit system for new satellite networks to discourage paper satellites, with all or part of the deposit returnable when the system entered into service (ITU Doc. CMR97/8-E 1997, pp. 8–9).

As elaborated in Chap. 6, these financial due diligence approaches did not have a broad level of support as did the administrative approach going into the conference. Their consideration was made the more complex by the Council's decision scant months before the WRC to impose cost recovery charges on satellite network filings in order to strengthen the financial base of the union.

## WRC-97: Considerations and Outcomes

The US delegation's preparations for WRC-97's consideration of Resolution 18 were led by the Federal Communications Commission (FCC), the US independent telecommunications regulator, and they included robust participation by satellite industry experts and representatives of federal government satellite systems, including the Department of Defense and the National Aeronautics and Space Administration (NASA). The commercial interests sought to establish a regulatory approach that would actively address the problem of over-filing, including detailed administrative due diligence requirements in addition to several changes to the Radio Regulations governing the coordination process. Some of the federal government operators, however, were reticent to reveal too many details of their proposed networks and opposed additional ITU regulatory burdens being assessed against their operations. The US delegation was in complete agreement to oppose any financial due diligence measures (US Delegation to WRC-97 Report 1997).

The Europeans supported administrative due diligence, as well as other regulatory reforms, including reducing the 9-year period for bringing a satellite network into use and simplifying advance publication information. The European Common Proposals to the conference left the door open for consideration of financial due diligence following further evaluation (ITU Doc. CMR97/5-E 1997, pp. 18–24). Luxembourg and Norway, the notifying administrations for SES and Telenor, respectively, observed during the preparations for the conference that "there is no panacea or magic solution to the problem of 'paper satellites'" (SC97-2/3 1997, p. 2). Instead, there must be a series of incremental steps to eliminate the problems of orbital congestion and paper satellites. These steps include submission of due diligence information to confirm that the satellite network has been brought into use. The two nations also urged consideration of two financial due diligence approaches—deposits and annual registration fees.

Australia was also supportive of the administrative due diligence approach, but it strongly believed that such an approach must be combined with a financial due diligence in order to effectively handle the problem of paper satellites. To the Special Committee, Australia proposed as a way forward:

> [T]he procedural due diligence approach should be considered by WRC-97 for adoption together with annual registration fees, to be effective immediately; a part of this approach would comprise a WRC-97 Resolution incorporating "best practices" which administrations would be urged to incorporate into their domestic legislation to weed out paper satellites. The mechanism to introduce financial deposits would be prepared, but its implementation would be deferred pending experience in the application of procedural due diligence, filing fees and annual registration fees. The BR Director would be asked to report to WRC-99 on the results achieved in the intervening 2 years, on the understanding that if the situation does not indicate sufficient improvement by WRC-99, then the financial deposit approach should be considered at WRC-99 for adoption. The Plenipotentiary Conference in 1998 would be advised of this approach so that it could amend the Convention and Constitution if necessary to provide for the possible adoption of financial deposits by WRC-99. (ITU Doc. SC97-2/5 1997)

WRC-97's consideration of Resolution 18 proved to be particularly divisive. A conference working group was able to draft a new resolution on administrative due diligence, but it faltered when attempting to reach a similar agreement on financial due diligence. A conference plenary considered an interim report from the working group on Resolution 18 along with a last-minute proposal submitted jointly by Australia, Luxembourg, Norway, the Netherlands and Sweden that contained a draft resolution to impose annual coordination and registration fees on most satellite networks in addition to administrative due diligence measures (Australia, Luxembourg, Norway 1997). Mexico, supported by several nations (including Tonga, the United States, and Russia), opposed consideration of the late proposal. In an extremely rare procedural move, Mexico invoked the Conference Rules of Procedure to force a vote on a motion to close the debate on the approval of the agenda of the session without including the late proposal. Upon further debate, the meeting reached agreement to invite future plenipotentiary conferences to consider the issue of financial due diligence (ITU Doc. CMR97/249(Add 1)-E 1997).

Ultimately, WRC-97 adopted the administrative due diligence solution to address the problems of backlog and overfilling. The solution is contained in Resolution 49: "Administrative due diligence applicable to some satellite radiocommunication services," which requires disclosure of high level data to demonstrate the planned implementation of proposed satellite networks in the fixed-satellite service, mobile-satellite service, the broadcasting satellite service that are subject to coordination under specific provisions under Article 9 of the Radio Regulations (pertaining to most commercial communication satellites) and the for modifications or additions to the plans.

Under Resolution 49, an administration seeking to coordinate a satellite network or requesting recording in the MIFR, is required to send to the bureau as early as possible, and before bringing into use, the following due diligence information:

1. Identity of the satellite network
    (a) Identity of the satellite network
    (b) Name of the administration

   (c) Country symbol
   (d) Reference to the advance publication information (or to the request for modification of the plans in AP30/30A)
   (e) Reference to the request for coordination (not applicable for AP30/30A)
   (f) Frequency band(s)
   (g) Name of the operator
   (h) Name of the satellite
   (i) Orbital characteristics

2. Spacecraft manufacturer

   (a) Name of the spacecraft manufacturer
   (b) Date of execution of the contract
   (c) Contractual "delivery window"
   (d) Number of satellites procured

3. Launch services provider

   (a) Name of the launch vehicle provider
   (b) Date of execution of the contract
   (c) Anticipated launch or in-orbit delivery window
   (d) Name of the launch vehicle
   (e) Name and location of the launch facility (Final Acts WRC-97, Resolution 49, Annex 2, 1997)

Member states agreed that if complete due diligence information was not received by the required date, then the network filing should no longer be taken into account and should not be recorded in the MIFR. This would, in effect, strip the network of any priority over other filings and thus regulatory protection from harmful interference. Any provisional recording in the MIFR would be canceled. The Radio Regulations were amended to include these actions, although not without rigorous debate.

The administrative due diligence resolution amounted to a carefully crafted series of compromises accepted by consensus because it satisfied four principles that had guided its creation: "that the procedures should be equitable, taking into account the interests of both developing and developed countries; that they should not be overly bureaucratic; that they should be realistic and workable; and that they should take into account the systems submitted prior to the development of the new procedures." (ITU Doc. CMR97/334-E 1997, p. 3)

A US proposal to exempt government-owned and controlled satellite telecommunications services from the administrative due diligence procedure was not accepted by the conference. The United States took a reservation to the Final Acts of WRC-97 to state:

> The United States of America will make all reasonable efforts to comply with the administrative due diligence procedures contained in Resolution [49], but reserves the right not to in cases involving satellite networks or satellite systems that transmit government telecommunications as defined under No. 1014 of the Annex to the International Telecommunication Constitution. (Geneva, 1992) (ITU Doc. CMR97/400-E 1997, p. 23)

Russia took a similar reservation. Tonga reserved against any retroactive application of the decisions of WRC-97.

The 1997 conference did not approve the proposals of Australia, Luxembourg and others to adopt a system of financial due diligence. Mexico asserted that before the financial approach is considered, the results of the administrative due diligence approach should be evaluated, in addition to the impact of the recently enacted satellite network cost recovery process. Moreover, Mexico observed, "the Union was not a commercial entity and that administrations should not, therefore, pay it any fee for authorization to use certain parts of the spectrum or certain positions on the geostationary orbit." (ITU Doc. CMR97/389-E 1997, p. 7) This intervention was supported by Colombia, Tonga, the United States, Iran, Ecuador, Venezuela, Pakistan, Russia, Papua New Guinea, Bulgaria and Saudi Arabia on behalf of Arab countries, whereas the Czech Republic, Australia, Sweden, Norway, and Laos spoke in favor of such an approach. Japan stated that it feared that "financial measures were a remedy worse than the evil they were intended to cure since they might impede the development of genuine satellite networks," but that Japan could support further studies." (ITU Doc. CMR97/389-E 1997, p. 7) China, Singapore, Indonesia and Vietnam remarked that the time was not yet ripe for consideration of financial due diligence measures.

The chairman of the conference (Australia's Roger Smith) summarized the debate, observing that a majority of conference participants were opposed to the proposed resolution on financial due diligence, although many administrations thought that the subject should be explored in greater depth (ITU Doc. CMR97/389-E 1997, p. 8). Thus, financial due diligence was not accepted by the WRC-97. However, the conference did agree to include a provision in Resolution 49 to note that several years may be needed in order to assess whether the administrative due diligence measures produced satisfactory results. It further instructed the director to provide a report to the next World Radiocommunication Conference, and to future relevant conferences, on the results of these administrative due diligence procedures, thus keeping open the possibility of considering additional approaches to resolving the problems identified in Resolution 18.

WRC-97 made additional decisions responsive to Resolution 18 to improve the procedures governing satellite network filings. Perhaps most significant was its decision to shorten the regulatory period for bringing a satellite network into use. Previously, the regulatory period for bringing a satellite network into use was 9 years—a period of 6 years plus an automatic 3-year extension, from Advance Publication Information (API). With WRC-97's change, administrations would have only 5 years to coordinate and notify their satellite networks. They could receive a 2-year extension of the 5-year period only upon filing complete due diligence information *and* showing extenuating circumstances beyond their control to justify the extension, such as launch failure, problems with satellite design specifications, or *force majeure*.

Another innovation was to amend the Radio Regulations to require a coordination request to be filed within 2 years of the API submission. The conference also adopted Recommendation 36, "Role of international monitoring in reducing apparent congestion in the use of spectrum and orbit resources," encouraging the ITU-R to make studies on use of monitoring and administrations to provide monitoring facilities and to cooperate in joint monitoring programs in order to achieve more effective use of the geostationary satellite orbital and spectrum resources (ITU Final Acts of the World Radiocommunication Conference, Geneva 1997).

## Plenipotentiary Approval, 1998

Not long after the WRC-97, the plenipotentiary conference was convened in Minneapolis in 1998 where several administrations sought to force reconsideration of the issue of financial due diligence. The Europeans submitted a proposal calling for the plenipotentiary conference to rule, in principle, that World Radiocommunication conferences have the authority to decide on financial measures, including whether to institute financial due diligence into the Radio Regulations (ITU Doc. PP98/32-E 1998, p. 63). Luxembourg, the Netherlands and Sweden submitted a complementary proposal that, in furtherance of Resolution 18, the coordination and notification procedures, which are the foundation of the ITU's mandate in space, be continuously kept as current as possible. To achieve this, they proposed that every World Radiocommunication conference be charged to review and update the coordination and notification procedures "to ensure that they reflect the latest technologies" (ITU Doc. PP-98/97-3 1998).

Canada's proposal aligned with the Europeans, asserting that further action under Resolution 18 should include possible financial due diligence measures, such as filing fees, an annual registration fee, or a deposit system to address the overfilling issue should administrative due diligence prove to be ineffective (ITU Doc. PP-98/19-E 1998, p. 2). Canada's proposal also discussed the need to rebalance the union's underlying financial structure to reflect the growing role of the operators from the private sector and the need for enhancing their financial participation in the union's activities, over and above the annual payment required for sector membership.

Australia and 11 Asian countries (including China) proposed a new resolution establishing that WRCs are competent to adopt financial due diligence measures and adding an item to the agenda of the next World Radiocommunication Conference in 2000 (WRC-2000) consideration of the financial due diligence issue. In addition, the proposed resolution would instruct the director to develop proposals for WRC-2000 containing the regulatory and administrative framework to support financial deposits and/or annual registration fees for satellite network filings (ITU Doc. PP-98/70-E 1998). Malaysia, on the other hand, proposed that the plenipotentiary conference adopt a new resolution to study the impact of proposed financial due diligence measures on developing countries before taking any further steps to adopt such a process.

Tonga also proposed a new resolution to defer any consideration of financial due diligence until reviews are completed by the next two World Radiocommunication

Conferences. Tonga observed that Resolution 18 was having the unfortunate effect of actually worsening the problem, as administrations were increasing the rate of satellite network filings in order to avoid future regulatory constraints, and that these actions were destabilizing the ITU's regime. Moreover, Tonga raised the point that was to be greatly debated in the ensuing discussions—that the coordination procedures are a core part of the constitution:

The use of punitive financial measures as a means to discourage members from exerting their constitutional rights can only be seen as being in fundamental opposition to the basic tenets of the union itself (ITU Doc. PP-98/93-E 1998, p. 5).

In the end, the Minneapolis Plenipotentiary Conference adopted Resolution 85, "Evaluation of the administrative due diligence procedure for satellite networks adopted by the World Radiocommunication Conference (Geneva 1997)," which instructed WRC-2000 to evaluate the effectiveness of administrative due diligence. As part of the final compromise, Luxembourg agreed to remove a proposed provision regarding consideration of financial due diligence options. However, Luxembourg, the Netherlands, and Sweden did prevail on their proposal for continuous reform. The conference approved a new Resolution 86 on "Coordination and notification procedures for satellite networks" as a further follow on to Resolution 18. The resolution instructed WRC-2000 and subsequent world radio conferences to:

> [C]ontinually review and update the advance publication, coordination and notification procedures, including the associated technical characteristics, and the related Appendices of the Radio Regulations, so as to ensure that they reflect the latest technologies, as well as to achieve additional simplification and cost savings for the Radiocommunication Bureau and administrations. (Final Acts of the Plenipotentiary Conference, Minneapolis 1998, Resolution 86, p. 246)

With these actions, the conference decided that work under Resolution 18 (Kyoto 1994) was complete and therefore the famous resolution was abrogated. The Resolution 86 that it adopted still remains in effect and serves as the basis of WRC standing Agenda Item 7, as will be further discussed below.

## Later Developments

Later World Radiocommunication conferences and plenipotentiary conferences continued the study of due diligence as they had been tasked. At the next World Radiocommunication Conference in 2000, which was convened in Istanbul, the debate about financial due diligence continued and proved equally controversial. However, because administrations were making use of the maximum regulatory period for bringing their satellite networks into use, it would be until at least November 2003 until results of the administrative due diligence procedures could be observed and analyzed. Following what even the ITU later called a "long, drawn-out debate" (ITU Newsroom 2000), the Istanbul conference adopted Resolution 81,

"Evaluation of the administrative due diligence procedure for satellite networks" resolving:

1. That further experience is needed in the application of administrative due diligence procedures adopted by WRC-97, and that several years may be needed to see whether the procedure produces satisfactory results.
2. That it is premature to consider the adoption, among other procedures, of any financial due diligence procedures. (Final Acts of WRC-2000 2000, p. 419)

WRC-2000 further observed that the backlog and resulting processing delays had reached the point that administrations might have to wait as long as 3 years for their coordination requests to be published by the bureau, thus seriously shortening the window within which the network's coordination had to be completed and brought into use. The conference therefore adopted a series of temporary procedures "to ensure the continued viability and credibility of the ITU satellite coordination process" (Final Acts of WRC-2000, p. 379). The conference also mandated electronic submission of coordination filings (including advance publications, notifications), providing an extended transition period for developing countries filing less than three networks per year.

By 2002, the RRB, which had been tasked by the Council to develop rules of procedure to assist in resolving the backlog situation, concluded that "all possibilities of eliminating the backlog through application of the current regulatory framework have been exhausted and that within the current framework of the Radio Regulations no further action can be taken to successfully eliminate the backlog" (Chairman, RRB 2002, Doc. RRB02/331-E, p. 2).

The board adopted a series of rules of procedure, to be reviewed by the following World Radiocommunication Conference in 2003, which called for the elimination of certain regulatory examinations by the bureau and use of a coordination arc approach in fixed satellite service bands above 3 GHz that had been approved by WRC-2000. In addition, the Council created a satellite backlog action group to consider various approaches to backlog reduction; a software experts group to develop new software tools; and made additional resources available to the Bureau to staff the processing of the backlog of satellite network filings (Council Resolution 1182 2001, Doc. C2001/118-E).

The 2002 Plenipotentiary Conference, convened in Marrakesh, reviewed a report on administrative due diligence from the director of the Radiocommunication Bureau, but took no additional action save to declare PP-98 Resolution 85 as complete. It maintained and revised Resolution 86 to request World Radiocommunication conferences to continue to review and update the advance publication, coordination, notification and recording procedures for satellite network filings to facilitate equitable access and other aims of Article 44; to ensure that these procedures reflect the latest technologies; and to achieve simplification and cost savings (Final Acts of Plenipotentiary Conference (Marrakesh 2002, p. 257)). The continuing discussions on satellite network cost recovery continued to prove controversial.

At long last, the backlog situation showed improvement as the world returned to Geneva in the extremely hot summer of 2003 for the next World

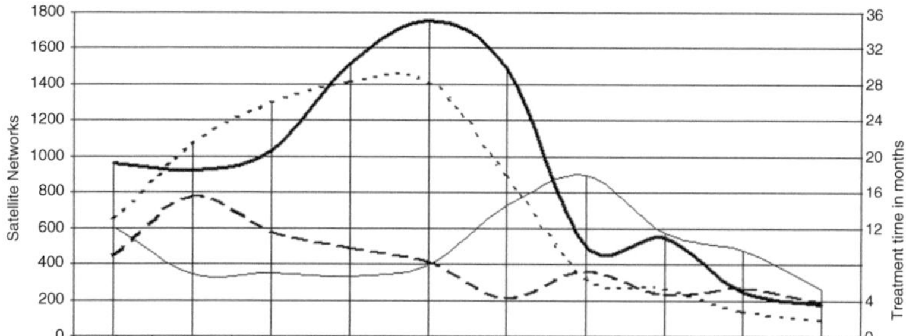


| | 1997 | 1998 | 1999 | 2000 | 2001 | 2002 | 2003 | 2004 | 2005 | 2006 |
|---|---|---|---|---|---|---|---|---|---|---|
| Filings Received | 447 | 773 | 576 | 489 | 416 | 208 | 365 | 229 | 268 | 190 |
| Filings Published | 605 | 346 | 350 | 328 | 394 | 728 | 898 | 570 | 475 | 258 |
| Under Treatment | 642 | 1069 | 1295 | 1410 | 1409 | 890 | 324 | 269 | 143 | 91 |
| Treatment Time (Months) | 19.2 | 18.4 | 20.6 | 30.3 | 35.1 | 29.8 | 10 | 10.9 | 5 | 3.6 |

**Fig. 5.2** Treatment time in the processing of requests for coordination (ITU Doc. CMR07/4 (Add. 1)-E, 2007, p. 3)

Radiocommunication Conference. The director of the Radiocommunication Bureau reported that the backlog was finally diminishing, due to the combined effects of:

- Reduction in the number of notices received.
- Application of cost recovery to new filings received.
- Mandatory electronic filing and publication of notices "as received" (as required by Resolution 55 (WRC-2000)).
- Introduction of new rules of procedure, approved by the RRB, relating to receivability of notices (validation for completeness) and temporary rules relating to examination pursuant to Nos. 9.35 and 9.36 of the Radio Regulations.
- Improved automation in internal processing in the bureau.
- Additional budget resources to enable recruitment of more engineers for examination of notices (approved by Council 2001). (ITU Doc. CMR03/4(Add.2)-E 2003, p. 1).

The 2003 conference updated the administrative due diligence resolution, Resolution 49. It also acted to revise the deadline for filing administrative due diligence information to align it with other changes it made to Article 11 of the Radio Regulations. The conference decided to streamline the regulations by eliminating the 2-year extension period to the 5-year regulatory period for bringing a satellite network into use, in favor of a single 7-year period applicable to all (ITU Final Acts of the World Radiocommunication Conference, Geneva 2003).

By the time of the 2007 World Radiocommunication Conference, the Director of the Radiocommunication Bureau noted that "drastic improvements" were becoming evident in the processing of satellite network filings due to:

- More staff resources, better software, and improved staff morale.
- Reduction of the number of notices submitted to the bureau, indicating that member states are no longer filing paper satellites, but only filings for their actual requirements.
- Application of cost recovery (addressed in the following section) (ITU Doc. CMR07/4(Add.1)-E 2007, p. 2).

This reduction is reflected in Fig. 5.2.

Thus, it was possible to observe that the administrative due diligence approach was actually working, perhaps in combination with other discrete regulatory reforms and other external factors. In any event, the call for financial due diligence was quieted.

## References

Australia, Bangladesh, Bhutan: Implementation of Changes to the ITU's Frequency Coordination and Planning Framework for Satellite Networks. ITU Doc. PP-98/70-E (1998)

Australia, Luxembourg, Norway: Proposals for the Work of the Conference, Geneva. ITU Doc. CMR97/249(Addendum 1)-E (1997)

Australia, Luxembourg, Norway: Proposals for the Work of the Conference, Geneva. ITU Doc. CMR97/249-E (1997)

Australia: Proposals for the Work of the Conferences, Kyoto. ITU Doc. PP-94/42-E (1994)

Australia: Resolution 18 (Kyoto)—Views and Proposals, Geneva. ITU Doc. SC97-2/5-E (1997)

Canada: Proposals for the Work of the Conference, Minneapolis. ITU Doc. PP98/19-E (1998)

Chairman of the Committee Four: Summary Record of the Tenth Meeting of Committee Four (Strategic Policy and Plans), Kyoto. ITU Doc. PP-94/231-E (1994)

Chairman of the Conference: Declarations, Geneva. ITU Doc. CMR97/400-E (1997)

Chairman of the Conference: Minutes of the Ninth Plenary Meeting, Geneva. ITU Doc. CMR97/389-E (1997)

Chairman of the Conference: Minutes of the Sixth Plenary Meeting, Geneva. ITU Doc. CMR97/334-E (1997)

Chairman, Radio Regulations Board: Some Considerations on the Rules of Procedure S9.35 and S9.36, Geneva. ITU Doc. RRB02/331-E (2002)

Council: Resolution 1182, Eliminating the backlog in the Radiocommunication Bureau's processing of satellite network filings, Geneva. ITU Doc. C2001/118-E (2001)

Department of State: US Delegation to WRC-97 Report, Washington (1997) (unpublished)

Director, Radiocommunication Bureau: Report on Resolution 18 (Kyoto, 1994), Geneva. ITU Doc. CMR97/8-E (1997)

Director, Radiocommunication Bureau: Processing of Satellite Network Filings and Related Issues, Geneva. ITU Doc. CMR03/4(Addendum 2)-E (2003)

Director, Radiocommunication Bureau: Activities of the Radiocommunication Sector in the Period between WRC-03 and WRC-07, Geneva. ITU Doc. CMR07/4(Addendum 1)-E (2007)

European Conference of Postal and Telecommunications Administrations: European Common Proposals for the Work of the Conference, Geneva. ITU Doc. CMR97/5-E (1997)

European Conference of Postal and Telecommunications Administrations: European Common Proposals for the Work of the Conference, Minneapolis. ITU Doc. PP98/32-E (1998)

International Telecommunication Union: Final Acts of the Plenipotentiary Conference, Kyoto. International Telecommunication Union, Geneva (1994)

International Telecommunication Union: Final Acts of the World Radiocommunication Conference, Geneva. International Telecommunication Union, Geneva (1995)

International Telecommunication Union: Final Acts of the World Radiocommunication Conference, Geneva. International Telecommunication Union, Geneva (1997)

International Telecommunication Union: Final Acts of the Plenipotentiary Conference, Minneapolis. International Telecommunication Union, Geneva (1998)

International Telecommunication Union: Final Acts of the Plenipotentiary Conference, Marrakesh. International Telecommunication Union, Geneva (2002)

International Telecommunication Union: Final Acts of the World Radiocommunication Conference, Geneva. International Telecommunication Union, Geneva (2003)

International Telecommunication Union, Newsroom: Main Results of WRC-2000. Geneva: International Telecommunication Union. http://www.itu.int/ITU-R/index.asp?category=conferences&rlink=wrc-00-results&lang=en (2000)

Luxembourg, Norway: Resolution 18: Due Diligence and Other Items from SC-4, Geneva. ITU Doc. SC/97-2/3-E (1997)

Luxembourg, Netherlands, Sweden: Proposals for the Work of the Conference, Minneapolis. ITU Doc. PP-98/97-E (1998)

Malaysia: Proposals for the Work of the Conference, Minneapolis. ITU Doc. PP-98/76-E (1998)

Tonga: Proposals for the Work of the Conference: Matters Related to the Implementation of Resolution 18, Minneapolis. ITU Doc. PP-98/93-E (1998)

# Chapter 6
# Satellite Network Cost Recovery

In the last chapters we have reviewed the development of administrative due diligence, and the rejection of financial due diligence, to address the overfilling of satellite networks. This chapter returns to 1994 to trace the ITU's parallel activity of developing and adopting a separate process that also affects satellite network filings: satellite network cost recovery. This regime was adopted primarily for the purpose of strengthening and expanding the financial base of the union. However, this action also ultimately served to contribute to the reduction of paper satellite filings and the backlog, as the proponents of financial due diligence proponents had originally intended (Fig. 6.1).

## Improving the Union's Financial Base, 1994

By the 1990s, the changes in the telecommunications marketplace were having repercussions on the health of the ITU's finances. The union was facing a plateau in the number of contributory units paid by member states on a voluntary basis to fund the union, while at the same time anticipating growing demands for services. The decline of member state contributions was the natural result of a trend occurring in many nations of separation of regulatory and operational functions as they privatized their public telecommunications networks. The regulator, then finding itself bereft of the accustomed operational revenues, would be faced with declining budgets, which ultimately translated into lower levels of participation and financial contribution to the union. The private operators, who have lesser rights of participation in the Union's activities because they are non-state members, failed to pick up an offsetting level of financial support to the union (ITU Doc. PP-98/13-E 1998, p. 2). A looming deficit was the result.

Thus, in addition to Resolution 18 to combat paper satellites, the 1994 Plenipotentiary Conference adopted Resolution 39, "Strengthening the financial base of the International Telecommunication Union." The Resolution, based again

A.L. Allison, *The ITU and Managing Satellite Orbital and Spectrum Resources in the 21st Century*, SpringerBriefs in Space Development,
DOI 10.1007/978-3-319-05314-1_6, © Audrey Allison 2014

**Fig. 6.1** US Vice President Al Gore opens the 1998 Plenipotentiary Conference in Minneapolis, Minnesota, USA (ITU File Photo)

upon a proposal from Australia, sought to improve the financial foundation of the union by encouraging wider financial participation by non-member states who benefit from ITU services. The proposal also discussed cost reduction options and finding ways to make better use of ITU informational resources, including, where appropriate, "charging fees for access to ITU services, particularly where these services are sought on a discretionary basis or to a greater extent than the level of facilities generally provided." (Final Acts of the Plenipotentiary Conference (Kyoto 1994), p. 200) The Resolution also recalled the ITU's recent experience in charging "cost recovery" fees for some products and services (such as publications) that were already being routinely charged, and should be considered along with other revenue options for other ITU activities in an effort to place the union on a stronger financial footing for the future. The resolution further charged the secretary-general to clearly identify the costs of activities and services provided by the ITU in order to determine options for implementation of this resolution for action by the Council.

## Council Actions, 1997–1998

During the years following Kyoto and leading up to the next plenipotentiary conference in Minneapolis in 1998, the Council explored the idea of charging full cost recovery fees for the services provided by its Radiocommunication Bureau staff in processing and publishing satellite network filings and in maintaining registrations in the Master International Frequency Register (MIFR). In 1997, the Council adopted Resolution 1113 (1997), "Cost recovery for the processing by the Radiocommunication Bureau of space notifications." It instructed the secretary-general to develop fee schedules for processing satellite network filings based on full cost allocation and to include the proposed income and expenditure in future budgets of the union.

At its 1998 session in advance of the plenipotentiary conference, the Council moved forward with the implementation of cost recovery for satellite network filings with Decision 480, "Implementation of processing charges for satellite network filings and administrative procedures." The Council noted in the decision that "cost recovery is to be considered also as an incentive to reduce the number of 'paper satellites' as requested in Resolution 18." (Kyoto 1994) (Council Decision 480 1998) The Council determined that the processing charge per page would be calculated on the basis of the average number of pages in the corresponding Radiocommunication Bureau publication needed to inform administrations of that filing. The cost per page would be determined for advance publication, coordination requests, and requests for modification of the plans. The Council also considered the issue of whether payment could be made directly to the ITU by the satellite operator, including private satellite operators, rather than by the notifying the administration. This was of particular concern to regulators of major commercial satellite operators, such as the U. S. Federal Communications Commission.

The councilors also discussed whether to limit these fees to commercial satellite systems, and whether to exempt developing countries from them. In the end, the Council reached a consensus to extend cost recovery to all satellite networks and to allow each member state one "free filing" per year, up to 7,100 pages. Since developing countries would not normally submit more than one filing in a single year, they would, in effect, be exempted. The 1998 plenipotentiary conference was invited to provide guidance to the Council as to the dates of application of the proposed fee schedules. Many Council members sought earlier dates in order to promote equitable access to spectrum and orbits by "real" satellite networks, in addition to strengthening the financial base of the union, as called for by Resolutions 18 and 39 of Kyoto (ITU Doc. PP98/67-E 1998, pp. 1–2).

## Minneapolis Adopts Cost Recovery, 1998

The 1998 plenipotentiary conference in Minneapolis provided a forum for continuation of the discussions on both topics of satellite network cost recovery and paper satellites. As detailed above, much of the focus of the discussions in Minneapolis concerned financial due diligence, rather than cost recovery. On the latter issue, some administrations continued to question the appropriateness of charging for the processing of satellite network filings, which, after all, is a core service provided by the ITU in response to treaty obligations on member states. The United States continued to seek an exemption for non-profit services, such as the Global Positioning

System (GPS) and meteorological satellites that provide services for the benefit of the entire world community. Others noted that this concern should be diminished due to the fact that such systems have lesser coordination requirements leading to the assessment of smaller cost recovery fees. Additionally, an administration could apply its one free filing per year to such a non-commercial system (ITU Doc. PP-98/292, 1998, pp. 12–13).

Many conferees expressed a broader concern that the "cost recovery" activity could easily lead to efforts to make a profit (referred to as "revenue generation") from satellite network cost recovery fees, perhaps using such proceeds to subsidize other ITU activities wholly unrelated to processing of satellite network filings. Finally, Iran, Malaysia and Tonga jointly proposed that satellite network cost recovery be postponed until studies could be undertaken on its potential impact on developing countries (ITU Doc. PP-98/240-E 1998).

The conference followed-up on the Council's request for a date of implementation for its recent Decision 480. It had been unable to agree on the date of application of the satellite network cost recovery processing charges. A key point was whether the date should be retroactive to 1997, when the Council had adopted the original decision to apply cost recovery to satellite network filings. Another option would have been 1998, when the Council adopted Decision 480. Or another option would have been 1999, which would have avoided retroactive application. There was concern that adopting a date in the future might serve to encourage a rush of additional filings in advance of the deadline, thus increasing the backlog—clearly a counterproductive result. The conference ultimately decided that the cost recovery procedures would apply to filings filed on or after June 27, 1997. This decision was reflected in Resolution 88 (Minneapolis 1998), "Implementation of processing charges for satellite network filings and administrative procedures." (Final Acts of the Plenipotentiary Conference (Minneapolis 1998), pp. 248–49)

In furtherance of the Kyoto resolutions on expanding the financial base of the union, the conference also adopted a resolution providing guidance on cost recovery activities in general. Resolution 91 (Minneapolis 1998), "Cost recovery for some ITU products and services," endorsed cost recovery charges for some products and services covering only the exact cost of providing the product or service and not for generating revenue or profit. Moreover, it directed that the Council consider the following factors when deciding whether to apply cost recovery to a product or service:

1. When a product or service is provided for the benefit of a limited number of member states or sector members.
2. When a product or service is requested to a significantly greater extent by a small number of users.
3. When products or services are provided on a discretionary basis (Final Acts of the Plenipotentiary Conference, Minneapolis 1998, p. 254).

The developing countries, particularly Morocco, emphasized that such cost recovery was appropriate in cases where a service or product is primarily utilized by commercial entities rather than governments, and when it is requested to an extent greater than the level of facilities generally provided by at least an order of magnitude (ITU Doc. PP-98/51-E 1998, p. 4).

The conference amended No. 484 of the ITU Convention to provide the Council the authority to determine the criteria for application of cost recovery for some products and services of the union (Final Acts of the Plenipotentiary Conference, Minneapolis 1998, p. 78). The ITU's internal financial regulations were also amended to include application of cost recovery, along with guidance on applicability, cost allocation methodology and schedule of charges (ITU Doc. PP-98/51-E).

At the Extraordinary Session of Council convened immediately upon the plenipotentiary's conclusion, the councilors adopted a decision to establish a joint working group, including the participation of satellite network operators, to develop cost methodologies and a schedule of charges consistent with the outcomes of the conference.

In 1999, the Council adopted its landmark cost recovery action, Decision 482, "Implementation of cost recovery of satellite networks" containing a schedule of processing charges and a methodology to be reviewed periodically by the Council. The charging mechanism for satellite network filings was based on a flat fee charged for each category of filing plus an additional charge per page for filings going beyond a set maximum when published by the bureau in its weekly circular publication. For example, the flat fee for a coordination request for a typical geostationary satellite network was set at 17,500 CHF (approximately US$ 16,600). The Council invited the next World Radiocommunication Conference to consider modifications to the Radio Regulations to take this decision into account, including consequences of non-payment (Council Decision 482. 1999, ITU Doc. C99/94-E). The current version of Decision 482 is found in Appendix C at the end of this book.

## WRC-2000 Effort to Implement Cost Recovery

The 2000 World Radiocommunication Conference, convened in Istanbul, was faced with what proved to be the very challenging task of implementing the decisions taken by the plenipotentiary conference and Council on satellite network cost recovery into the Radio Regulations. It proposed to add the following footnote to the provisions of Article 9 on advance publication, Article 11, on coordination and notification, and relevant provisions of the plans to state:

> If the payments are not received in accordance with the provisions of Council Decision 482, as amended, on the implementation of cost recovery for satellite network filings, the Bureau shall cancel the publication, after informing the administration concerned. The Bureau shall inform all administrations of such action, and that the network specified in the publication in question no longer has to be taken into consideration by the Bureau and other administrations. The Bureau shall send a reminder to the notifying administration, not later than 60 days prior to due date of the payment if payment has not been received by that date. This provision was identified in reply to Resolution 88 (Minneapolis 1998) of the Plenipotentiary Conference (Minneapolis 1998) and shall enter into force at a date to be determined by the forthcoming plenipotentiary conference. (WRC-2000 Final Acts 2000, p. 425)

Some member states, including Colombia and Argentina, expressed strong reservations to this proposed provision, asserting that cancellation of a member state's satellite network filing was an infringement of its sovereign rights defined in the ITU Constitution and would result in an inequitable situation occurring between member states' satellite filings. Syria, on behalf of the Arab states, averred that this action based upon financial considerations was beyond the authority of a World Radiocommunication Conference and could only be taken by a plenipotentiary conference. WRC-2000 ultimately adopted the footnotes to the Radio Regulations shown above, but it was unable to resolve the matter of the date of implementation, which matter was left in abeyance pending action by the next plenipotentiary conference in 2002. This decision was reflected in Resolution 83, "Administrative procedures for cost recovery for satellite network filing."

## Cost Recovery Issues and Solutions

As the Radiocommunication Bureau began to apply the cost recovery process developed by the Council in Decision 482 to actual satellite filings, many difficulties arose, including cases of extremely large fees being assessed against some satellite operators. Because of these difficulties in implementation of the decision, the Council established a special *ad hoc* group to consider developing an alternative basis for charging for satellite network filings. An additional concern was that the system was not properly gauged to encompass only the bureau's actual costs of processing the filings (ITU Doc. C02/87-E 2002). By having the studies occur in an *ad hoc* group of the Council, private sector satellite operators who were sector members of the union were able to directly participate in the discussions to forge potential solutions for the Council's consideration.

The 2002 Plenipotentiary Conference, convened in Marrakesh, took action to recognize that WRC-2000s actions on cost recovery were appropriately taken in furtherance of Resolution 88 (Minneapolis 1998). The conference decided on an entry into force date for the new provisions of the Radio Regulations, implementing consequences of non-payment of cost recovery fees by August 1, 2003. It further instructed the Council to establish a new working group, open to sector members, to make recommendations on modifying Decision 482; to include in satellite network cost recovery only the "identifiable and auditable costs incurred directly in the processing of satellite network filings"; and to clarify the meaning of "actual costs" as required by Resolution 91 (Minneapolis 1998). But while this decision sought to ensure that cost recovery was indeed limited to collecting amounts representing the actual costs of the activity, the decision also gave the Council the authority to extend satellite network cost recovery processing beyond charging for advance publication and coordination requests, to include charges for the notification stage of the process.

For the next several years, the Council and its working groups considered various modifications to the cost recovery methodology, including approaches to accurately determine the actual costs of the bureau's provision of satellite network filing processing services, such as conducting time surveys of the bureau staff, and determining which of the general secretariat's costs should be included in the cost recovery calculus. During this time, the number of satellite filings and the amount collected from cost recovery fees declined, as many commercial satellite operators faced financial difficulties, including a number of high profile bankruptcies. Indeed, the amount of revenue that the ITU expected to gain from satellite network cost recovery, as stated in its financial plan for 2004–2007, was not being realized. There was a growing suspicion that the rates being recommended by the bureau for satellite network cost recovery fees were being calculated on the basis of meeting the amount in the budget line, rather than reflecting the actual costs of processing the filings. As the United States observed in a contribution to the ITU Council, "cost recovery" is not supposed to be "budget recovery." (ITU Doc. C05/52 2005, p. 2)

There were several iterations of Decision 482 during these years. When WRC-2000 decided to replace the paper publication of the weekly circular with an electronic publication, the ITU Council had to revisit its page-based fee calculation method and replace it with a new mechanism based on the number of "frequency units." This unit-based approach considered the size and complexity of the satellite network filing and was intended to more accurately represent the level of effort required by bureau staff. However, the eventual implementation of this new methodology resulted in additional anomalies; with some networks charged excessive fees. The Council later acted to reduce the fees charged to some large networks in cases in which the fees charged under the new methodology were dramatically higher than they would have been under the page methodology (Decision 513, Cost Recovery for Satellite Network Filings, ITU Doc. C03/88-E, 2003).

In 2005, Australia proposed to the Council a fee cap when five of its networks were charged cost recovery fees in excess of CHF 100,000 each (ITU Doc. C05/42-E 2005). The United Kingdom also complained of excessive fee invoices of CHF 490,127 and CHF 915,067 for two of its networks. (Note: 1 Swiss franc is worth about $0.91 US) It remarked, "[C]learly these sums are orders of magnitude greater than the Radiocommunication Bureau's costs associated in the … processing of these networks." (ITU Doc. C05/54-E 2005, p. 2) On the other hand, it was thought replacing this consumption-based system with a simpler flat fee approach for all satellite networks would not be an acceptable solution because smaller networks would, in effect, subsidize the larger ones.

The ITU Council ultimately decided that a "coherent, stable, and reliable mechanism" was needed to resolve these enduring problems with calculating satellite network cost recovery fees (ITU Doc. C05/82-E 2005, pp. 2–3). In 2005, it approved a new schedule of processing charges developed by the *ad hoc* group on cost recovery for satellite network filings. The new method provided an optional flat fee that would accommodate smaller networks and a unit charge for networks having more than 100 units. The Council also approved several "corrective measures," amounting to invoiced fee reductions of CHF 4.784 million (or about $4.5 million US), thus causing the Council to make a corresponding withdrawal from the ITU's reserve account to make up the difference (ITU Doc. C05/72-E 2005).

The Council's 2005 modification to satellite network cost recovery schedule of charges and methodology contained in Decision 482 worked well upon implementation and remains in effect today with only minor updates to reflect changes in the Radio Regulations by later WRCs. The painful multi-year trial-and-error process (only partially recounted here) did eventually develop a fair methodology. This led to its eventual wide acceptance by the ITU's larger community of space-faring nations, developing countries and satellite operators themselves. This end result and the troublesome years in between explains the great reluctance of the ITU and its members to reopen the satellite network cost recovery debate to this day.

## References

Australia: Additional Corrective Measures Relating to Cost Recovery on Satellite Network Filings for those Satellite Networks Subject to Decision 513. ITU Doc. C05/42 (Part II)-E (2005)

Chairman of the Conference: Incorporation of Cost Recovery in the Financial Regulations of the Union, Minneapolis. ITU Doc. PP-98/51-E (1998)

Chairman of the Conference: Minutes of the Twenty-Fourth Plenary Meeting, Minneapolis. ITU Doc. PP-98/292-E (1998)

Council: Resolution 1113, Cost Recovery for the Processing by the Radiocommunication Bureau of Space Notifications, Geneva. ITU Doc. C97/133-E (1997)

Council: Final Acts of the Plenipotentiary Conference, Kyoto. International Telecommunication Union, Geneva (1994)

Council: Final Acts of the Plenipotentiary Conference, Minneapolis. International Telecommunication Union, Geneva (1998)

Council Decision 513, Cost recovery for satellite network filings, Geneva. ITU Doc. C03/88-E (2003)

Council: Decision 480, Implementation of Processing Charges for Satellite Network Filings and Administrative Procedures, Geneva. ITU Doc. C98/128-E (1998)

Council: Decision 481, Establishment of a Joint Working Group on the Implementation of f Processing Charges for Satellite Network Filings and Administrative Procedures, Minneapolis. ITU Doc. C99EX/1IRev.1)-E (1998)

Council: Decision 482 (modified 2005), Implementation of cost recovery for satellite network filings. ITU Doc. C05/105-E (2005)

Council: Decision 482 Modified, Implementation of Cost Recovery for Satellite Network Filings, Geneva. ITU Doc. C02/68-E (2002)

Council: Decision 482, Implementation of Cost Recovery for Satellite Network Filings, Geneva. ITU Doc. C99/94-E (1999)

Council: Decision 501, Cost recovery on satellite network filings, Geneva. ITU Doc. C02/87-E (2002)

Council: Report by the Chairman of the Standing Committee on Finance to the Plenary Meeting. ITU Doc. C05/72-E (2005)

Council: Report from the Ad-hoc Group on Cost Recovery for Satellite Network Filings, Geneva. ITU Doc. C02/99-E (2002)

Council: Summary of the Seventh Meeting of the Standing Committee on Finance, Geneva. ITU Doc. C05/82-E (2005)

International Telecommunication Union: Final Acts of the World Radiocommunication Conference, Istanbul. International Telecommunication Union, Geneva (2000)

International Telecommunication Union: Final Acts of the Plenipotentiary Conference, Marrakesh. International Telecommunication Union, Geneva (2002)

Iran, Malaysia, Tonga: Implementation of Cost Recovery for Satellite Networks, Minneapolis. ITU Doc. PP-98/240-E (1998)

Secretary-General: Implementation of Processing Charges for Satellite Network Filings, Minneapolis. ITU Doc. PP-98/67-E (1998)

Secretary-General: Report on the Implementation of Resolutions 18 and 39 (Kyoto, 1994). Minneapolis. ITU Doc. PP-98/13-E (1998)

United Kingdom: Satellite Network Filing Cost Recovery: Categories 7 and 8, Geneva. ITU Doc. C05/54-E (2005)
United States of America: Satellite Network Cost Recovery, Geneva. ITU Doc. C05/52-E (2005)

## *Website*

ITU website for cost recovery for satellite network filings: http://www.itu.int/ITU-R/go/space-cost-recovery/en

# Chapter 7
# The Current State of Virtual Satellites

Twenty years after Resolution 18 and the paper satellites saga, administrative due diligence and satellite network cost recovery continue to be applied in much the same form as they were developed in the late 1990s and early 2000s. They are both success stories, yet both subjects continue to be raised as issues, most recently at the 2010 plenipotentiary conference in Guadalajara, and the 2012 World Radiocommunication Conference (WRC-12) in Geneva. Although the satellite network filing backlog is today essentially nonexistent, the number of satellite network filings submitted and the apparent congestion of the GSO do remain a challenge. In other words, we find that the phenomenon of "virtual satellites" persists (Fig. 7.1).

At WRC-12, the director of the Radiocommunication Bureau reported that "efficient use of the spectrum/orbit resource is one of the most crucial challenges facing the international community in efforts to promote worldwide telecommunication development." This challenge remains: "the ability to continue carrying out the vital work of recording frequency assignments in the [MIFR], while ensuring that frequencies and orbital positions are used in a rational, equitable, efficient and economic way." (ITU Doc. CMR12/4 (Add.7) 2012, p. 1)

Indeed, in 2011, the ITU-R had before it more than 3,300 filings for advance publication, coordination and notification of geostationary satellite networks in the unplanned bands submitted by 65 administrations. Of these, major filing administrations included (Table 7.1):

The number of filings for advance publication information, coordination and notification and the demand for access to the geostationary orbit have continued to increase year after year, despite the application of administrative due diligence and satellite network cost recovery, and the dip in filings seen in 2002, as illustrated in Figs. 7.2 and 7.3, below.

With filings continuing at this pace, the filers' ability to complete required coordination agreements within the requisite 7-year period becomes increasingly difficult. A new trend has thus been emerging. A growing number of notifications are being submitted to the bureau under No. 11.41 of the Radio Regulations and recorded provisionally in the MIFR without having completed required coordination.

A.L. Allison, *The ITU and Managing Satellite Orbital and Spectrum Resources in the 21st Century*, SpringerBriefs in Space Development,
DOI 10.1007/978-3-319-05314-1_7, 

**Fig. 7.1** The ITU's 2012 World Radiocommunication Conference opens in Geneva (ITU File Photo)

**Table 7.1** Excerpt of ITU-R count of unique GSO satellite network filings per administration (ITU-R Annual Space Report to STS-12 2011, pp. 97–98)

| Administration | Advance publication | Coordination | Notification | Total |
|---|---|---|---|---|
| China | 118 | 58 | 94 | 270 |
| France | 220 | 164 | 130 | 514 |
| UK | 89 | 106 | 45 | 240 |
| Japan | 55 | 51 | 99 | 205 |
| Luxembourg | 87 | 103 | 8 | 198 |
| Russia | 31 | 45 | 187 | 263 |
| UAE | 153 | 45 | 8 | 203 |
| USA | 76 | 144 | 448 | 668 |

At that time, Number 11.41 of the Radio Regulations provided:

> After a notice is returned under No. **11.38** [for failure to complete needed coordination], should the notifying administration resubmit the notice and insist upon its reconsideration, the Bureau shall enter the assignment provisionally in the Master Register with an indication of those administrations whose assignments were the basis of the unfavourable finding19. The entry shall be changed from provisional to definitive recording in the Master Register only if the Bureau is informed that the new assignment has been in use, together with the assignment which was the basis for the unfavourable finding, for at least four months without any complaint of harmful interference being made (see Nos. **11.47** and **11.49**). (ITU Radio Regulations 2008 ed.)

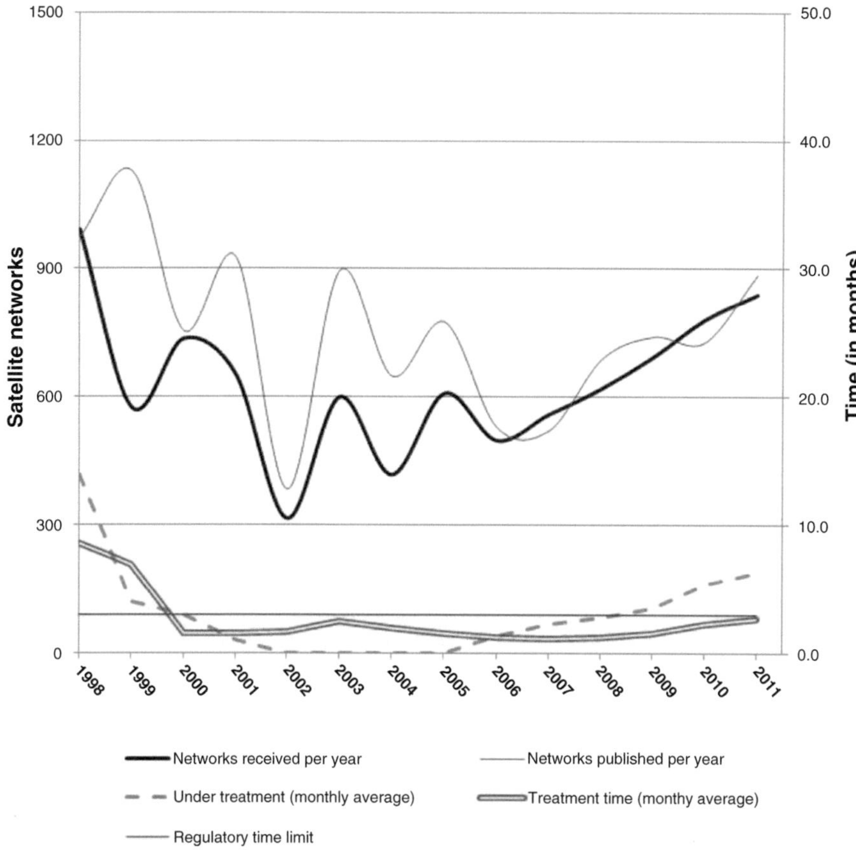


**Fig. 7.2** Advance publication of information (API) for satellite networks (Trends from 01.01.1998) (ITU, 2012)

In preparation for WRC-12, the bureau undertook an analysis of space frequency assignments recorded in the MIFR over the last 10 years to discern trends in filing and the validity of the information in the bureau's databases. The bureau also tracked the number of networks brought into use under No. 11.41 (provisional recording). These results were reported to WRC-12:

Stunningly, almost 100 % of notifications for satellite networks that were brought into use in 2012 were recorded on a provisional basis under No. 11.41 of the Radio Regulations, perhaps a logical outcome of the large number of coordination requests outstanding and the reported congestion in the GSO. In his report to the conference, the director explained that the peak in submissions in the year 2005, shown in Figs. 7.4 and 7.5 above, was the result of two main factors: the end of the regulatory period for bringing into use (seven or nine years) the glut of "paper satellite" filings during the period 1995–1999; and the entry into force of the Council's original satellite network cost recovery regime, Decision 482, in 1999 (ITU Doc. CMR12/4 (Add.7)(Add.1) 2012, p. 3).

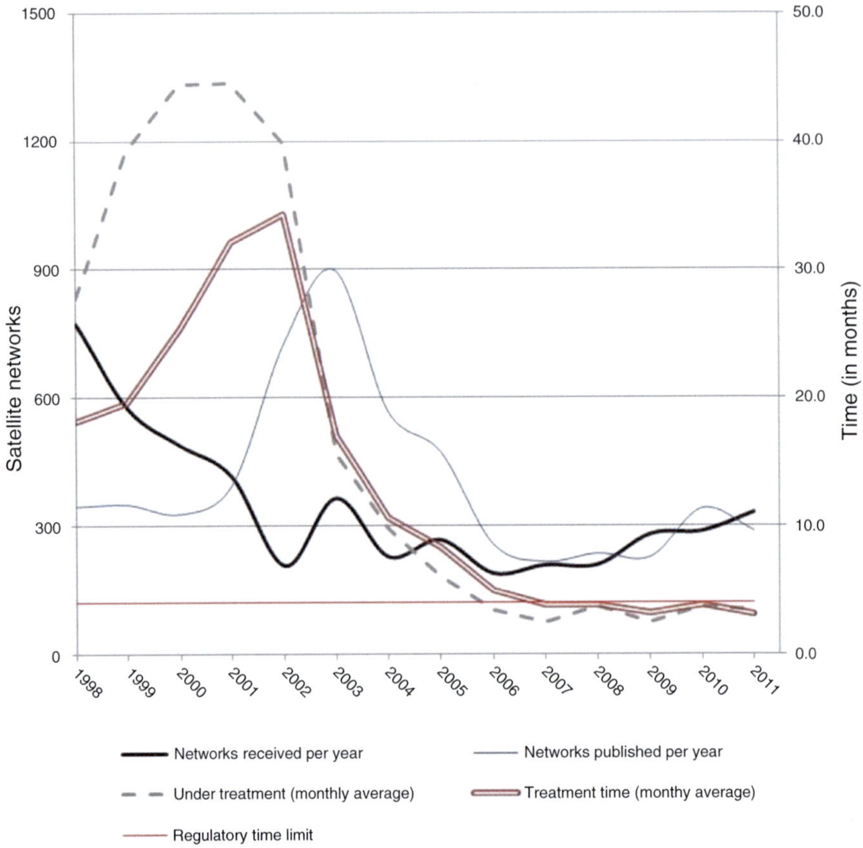


**Fig. 7.3** Coordination of requests for satellite networks (Trends from 01.01.1998) (ITU, 2012)

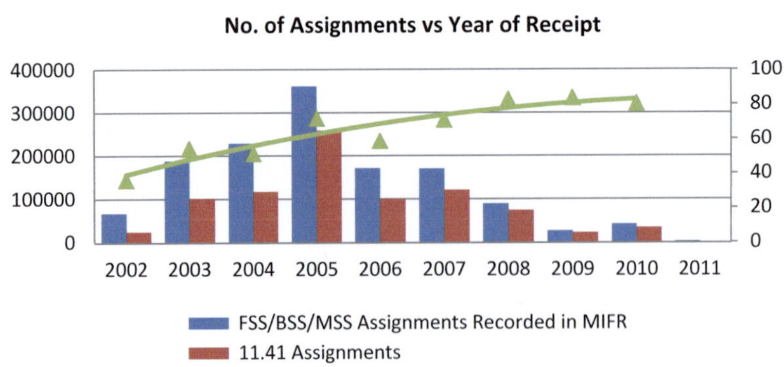


**Fig. 7.4** Number of frequency assignments for notification and recording in the MIFR per year (ITU Doc. CMR12/4 (Add.7) (Add.1), 2012, pp. 2–3)

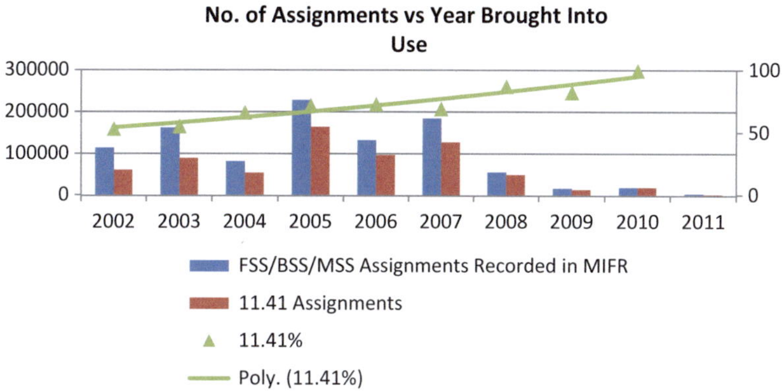


**Fig. 7.5** Number of notified assignments brought into use per year (ITU Doc. CMR12/4 (Add.7) (Add.), 2012, pp. 2–3)

Moreover, the growing number of coordination requests serves to compound the difficulties of achieving complete coordination for any single satellite network. The Bureau observed that the average number of administrations for coordination of GSO networks in the unplanned bands (under No. 9.7) had doubled since 2002 and that the number of satellite networks to be coordinated with increased by 67 % during that same timeframe (ITU Doc. CMR12/4 (Add.7) (Add.1), 2012). The director further noted that despite this failure across the board to complete coordination there had not been an increase in the number of complaints of harmful interference, as one might logically expect. This could be due to a number of factors, from increased efficiency in antenna design and modulation techniques to an absence of actual functioning spacecraft in these orbital positions (i.e., virtual satellites) (ITU Doc. CMR12/4 (Add.7) (Add.1), 2012). The satellite industry itself did not appear to be unduly concerned about this development.

## WRC-12: Satellite Regulatory Mini-Conference

During 4 ice-cold weeks in early 2012, more than 3,000 delegates representing 165 countries and a hundred other observers (including satellite operators) gathered in Geneva to consider changes to the international Radio Regulations governing satellites and other radio services and the use of spectrum and associated orbital resources. The WRC-12 conference considered an agenda of more than 30 items. One of these, Agenda Item 7, proved to be almost a conference unto itself.

Agenda Item 7 is a "standing" agenda item, meaning that it appears on every WRC agenda per the instruction of Plenipotentiary Conference Resolution 86 (Rev. Marrakesh 2002) discussed in Chap. 5. Agenda Item 7 tasks every World Radiocommunication Conference to consider possible changes in response to Resolution 86 (Rev. Marrakesh, 2002) of the Plenipotentiary Conference and "advance publication, coordination, notification and recording procedures for frequency assignments pertaining to satellite networks," in accordance with Resolution 86 (Rev. WRC-07) (ITU Doc. CMR12/1, 2012, p. 4).

WRC-12 considered more than 600 proposals under Agenda Item 7, which were organized into 34 discrete topics addressing, among other things, improvements to the provisions of Articles 9 and 11 and Appendix 4 (data elements) of the Radio Regulations, and parallel provisions of the satellite allotment plans in Appendices 30, 30A and 30B. Among the issues dealt with were: (1) improvements to Resolution 49 (i.e., Administrative Due Diligence); (2) clarifying the meaning of "date of bringing into use" and "regular operation" of a satellite network; (3) updating the provisions on suspension; (4) the regulatory impact of satellite leasing; and (5) extending the regulatory period for hosted payloads in the event of a launch failure. The bureau's role in maintaining the MIFR was also addressed under this agenda item. These topics were all intended to tighten the existing procedures. These items included: (a) reduction of the coordination arc; (b) creating a definitive list of coordination requirements; and (c) eliminating the 6-month period between advance publication and filing of the coordination request.

Although equitable access is a theme underlying consideration of Agenda Item 7, it is also the subject of its own standing WRC agenda item, Agenda Item 8.1.3, which calls for "action in response to Resolution 80 (Rev.WRC-07)." This topic is further addressed below.

## WRC-12 Actions on Administrative Due Diligence

In the preparations leading up to WRC-12, administrative due diligence and the continuing challenge of coordinating real satellites in an apparently congested environment led to discussions of tightening the coordination process and strengthening administrative due diligence. The ITU-R's Special Committee on Regulatory/Procedural Matters (Special Committee 2011) debated Resolution 49 in late 2010 as it prepared a report on regulatory issues for incorporation into the Conference Preparatory Meeting (CPM) to assist administrations in their preparation for WRC-12.

The United Kingdom proposed to the special committee that WRC-12 include in its consideration of Agenda Item 7 several topics that had been discussed in a series of workshops convened by the Radiocommunication Bureau on the subject of efficient use of the orbit/spectrum resource. The United Kingdom requested consideration of defining the date of bringing into use to address the practice under which some satellite operators drift a satellite briefly into one or multiple orbital positions and declare it or them as having been brought into use within the regulatory period required under the Radio Regulations (a practice later referred to as satellite "drifting" or "hopping"). The United Kingdom also recommended clarifying the period that a network can be suspended defined in Number 11.49 of the Radio Regulations. Finally, it proposed strengthening the administrative due diligence process by requiring the notifying administration to submit to the ITU evidence to substantiate a satellite's launch and operation, which would then be published by the bureau. In the future, when the satellite was replaced, the notifying administration would be required to update the due diligence information, so that the MIFR could be updated to accurately reflect the current situation with respect to that orbital position. These changes could be contained in a modification to Resolution 49 or in a new WRC resolution (ITU Doc. SC/33-E, 2010).

France also proposed reinforcing the Radio Regulations along similar lines. It tabled a proposed new resolution to define the date of "bringing into use." Furthermore, it proposed to replace Resolution 49 with a new resolution on "Information about the use of frequency assignments and systems in certain radio-communication services and frequency bands" (ITU Doc. SC/43-E 2010, p. 7). This resolution would require for systems in the commercial satellite bands (C, Ku and Ka-bands) due diligence information be submitted following the first time of being brought into use and also upon resumption of service following a suspension. Moreover, all existing networks in these frequency bands would be required to update their due diligence filings and to inform the bureau at the end of a satellite's life. France also proposed to update the required due diligence to include an "ITU ID number" assigned by the Radiocommunication Bureau for each spacecraft.

The United States intervened to state that it had difficulties with these proposals and that its position was that no changes should be made to Resolution 49.

The special committee's report to the CPM observed that administrative due diligence information filed by satellite network notifying administrations in response to Resolution 49 provides valuable information. This is a simple practice that has "proven effective in clearing paper satellite filings." (ITU Doc. CPM 11-2/2-E, 2011, p. 125) Resolution 49 has been extensively discussed at every WRC since its introduction in 1997. However, a weakness in this approach is that the due diligence information is filed prospectively at an early stage in the coordination process, and it describes the *plans* for manufacture and launch of a proposed satellite. There is no requirement to confirm this information or to update it following the scheduled implementation of the network.

Despite several proposals for change, and many complaints about its deficiencies, very few changes to Resolution 49 have been made at the WRCs since 1997. The special committee agreed that the practice of administrative due diligence "should be continued as a means to reflect the real utilization of spectrum and satellite orbital resources and to eliminate those recorded frequency assignments that were not actually brought into use." (ITU Doc. CPM11-2/2-E, 2011, Annex 18) It encouraged administrations to continue to consider the matter during their preparations for the conference.

The United States submitted a contribution to the conference preparatory meeting in early 2011 to supplement the special committee's treatment of the Resolution 49 issue under Agenda Item 7 to add consideration of the option of "no change" to the existing text of Resolution 49 at WRC-12. The United States noted that "[t]he current procedures for due diligence are adequate, and there is no need for additional regulatory requirements" (ITU Doc. CPM11-2/150-E, 2011, p. 3).

The CPM Report added the no change option and concluded:

> While fully appreciating the problem of 'virtual satellites' and the impact they have on administrations' ability to access to the orbit/spectrum resource and noting the sensitivity and history surrounding due diligence information, the following concerns should be considered with regard to modifying Resolution **49 (Rev. WRC-07)** or developing a new resolution addressing due diligence:

- Requiring more information from an administration or requiring an administration to provide updates would not improve accuracy nor seem to facilitate the implementation of the resolution;
- Introducing a requirement for "evidence" to substantiate the bringing of a satellite into use may create an atmosphere of mistrust between administrations and the BR whereas the ITU environment is based upon the word of an administration, good will, mutual respect, collaboration, and cooperation as given in Article 1 of the Constitution.

RR No. **13.6** provides a means to conduct consultations based on reliable information as to whether or not an administration has a satellite operating in accordance with the notified characteristics. The BR is making use of this provision in reconciling the MIFR, and the RRB is addressing the application of RR No **13.6** issue under Resolution 80 (Rev.WRC-07). That work is ongoing (ITU Doc. CMR12-3-E 2012, p. 240).

The US preparations for WRC-12, per its usual process, were performed on two separate tracks. One was the federal government's (civil and military) preparatory process, led by the Interdepartment Radio Advisory Committee, chaired by the Department of Commerce's National Telecommunication and Information Administration. The second track was led by the Federal Communications Commission (FCC 2011) on behalf of commercial and private operators, consumers and state and local governments. The FCC mainly acts through its own Federal Advisory Committee dedicated to WRC preparations. The FCC preparatory process for WRC-12 included the participation of major satellite operators, satellite manufacturers, and federal government observers. While both tracks develop draft WRC proposals, it is the US State Department that has the ultimate authority over the proposals that are finally submitted to the ITU (US Delegation Report World Radiocommunication Conference 12, 2012, pp. 19–22)

As had been the case in 1997, internal discussions in the United States on administrative due diligence were vigorous. The FCC Advisory Committee authored a proposal expressing the industry's view, shared by its regulator, that Resolution 49 should be updated and strengthened at WRC-12 to increase the transparency of the satellite coordination process and to reduce the number of virtual satellites. The industry observed that the increasing difficulty in obtaining access to the geostationary orbit was largely due to difficulties in completing coordination of new orbital locations and applying the provisions of the Radio Regulations. This problem was exacerbated by the fact that many unused spectrum and orbital resources are recorded in the MIFR and that this problem is most acute in the frequency bands where the actual usage is most congested (FCC Public Notice, DA 11–712, 2011, Attachment 1, pp. 18–19).

The problem could be alleviated by requiring only for those networks in the most severely congested commercial satellite bands the filing of additional Resolution 49 data after the launch of a satellite network to confirm that the data are correct and to ensure that the frequency assignments have been brought into use. The federal agencies continued to express concerns that such regulatory changes could be applied to government satellite operations in other frequency bands, if not now, then eventually–the slippery slope argument. The federal view was adopted in the inter-agency process, and the United States continued to propose no change to Resolution 49 to the WRC.

In his report to the conference, the director of the Radiocommunication Bureau also noted the lack of a regulatory requirement for administrations to update their due diligence information once their network is brought into use. One remedy would be to require such information within 30 days of bringing into use, or resuming use, of a frequency assignment to a satellite network, and to renew this information when there are other changes. An additional idea put forward by the bureau was extending Resolution 49 to require much more detailed information about satellite proposals, such as payload diagrams, transponder frequency plans, actual coverage information and power capabilities. Although it is not the role of the bureau to verify the accuracy of the information in due diligence filings, the information would be available to other interested parties to perform their own analysis and, if appropriate, to challenge the filing through bilateral discussions with the notifying administration with or without the assistance of the bureau (ITU Doc. CMR12/5 (Add.7)-E, 2012, p. 7).

The European administrations jointly submitted a proposal to WRC-12 "to establish a more robust and accurate system of due diligence." (ITU Doc. CMR12/5 (Add.28) (Add. 1)-E, 2012, p. 1). Europe "proposes to transform the current due diligence process from a situation where administrations declare what spacecraft is intended to be used to bring into use a satellite network to a situation where administrations declare what spacecraft has been used to bring into use a satellite network" (ITU Doc. CMR12/5 (Add.28) (Add. 1)-E, 2012, p.2). This revised process was proposed to be limited to networks in the frequency bands 6/4 GHz, 13-14/10-11-12 GHz and 30/20 GHz range.

Canada also proposed a modification to Resolution 49 in order to achieve additional transparency by requiring information following the bringing into use of a frequency assignment to a satellite network. The proposal, which was not limited to networks operating in any particular frequency bands, sought to add three Annexes to the Resolution to specify the elements of new due diligence filings to follow bringing into use, suspension, and bringing back into use (ITU Doc. CMR12/7[Add. 4]-E, 2012).

Similar to the Europeans, the Arab states jointly proposed to improve the accuracy of the due diligence process and to "reinforce the link between assignments to geostationary satellite networks and the spacecrafts used in certain frequency bands [which] would improve administrations' ability to access the radio-frequency spectrum and orbital resources," by modifying Resolution 49 and adding a new resolution to call for extended information for satellite networks operating in certain congested frequency bands (ITU Doc. CMR12/25(Add. 28) (Add.1)-E, 2012, p. 7). A difference was that the Arabs proposed to require spacecraft identification numbers, as the French has proposed to the special committee.

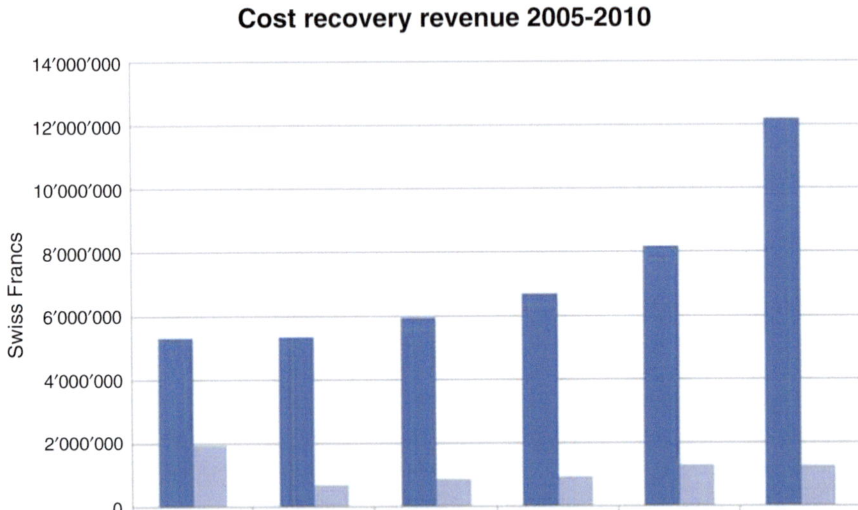


| | 2005 | 2006 | 2007 | 2008 | 2009 | 2010 |
|---|---|---|---|---|---|---|
| Revenue | 5'318'741 | 5'344'941 | 5'957'633 | 6'713'447 | 8'168'441 | 12'198'932 |
| Free Ent | 1'952'409 | 688'654 | 847'008 | 924'825 | 1'289'446 | 1'244'608 |

**Fig. 7.6** Cost recovery revenue 2005–2010 (ITU Doc. CMR12/4 (Add.1) (Rev.1), 2012, p. 15)

Several member states and regional groups proposed no change to Resolution 49, in agreement with the United States. Included were the Russian and Eastern European bloc (Regional Commonwealth in the Field of Communications), the Americas bloc (Inter-American Telecommunications Commission), Asia-Pacific (Asia-Pacific Telecommunity), and the administrations of Cameroon, Nigeria, Mexico, Cote d'Ivoire, Indonesia and Iran. The outcome of no substantive change to Resolution 49 emerged at WRC-12, while attention was focused on other challenges and improvements to the provisions of the Radio Regulations (Fig. 7.6).

## WRC-12 Actions on Satellite Network Cost Recovery

In 2011, the director of the Radiocommunication Bureau reported to his Radiocommunication Advisory Group (RAG), that:

> The adoption of Decision 482 on cost recovery for the processing of satellite network filings has definitely played its expected role by limiting, inter alia, coordination filings to projects which have a greater likelihood of becoming real. Through this decision, the problem of backlog in processing of satellite network filings which had accumulated since the early 1990's as a result of an incoming flow of "paper satellites" has been solved, and since the end of 2010, there is no backlog in any in any part of the BR process. (ITU Doc. RAG11-1/1(Rev-1)-E, 2011, pp. 1–2)

In addition, by 2011, 99 % of satellite network cost recovery invoices had been paid and the annual budgetary projections for satellite network cost recovery were exceeded. Thus, the director told the RAG, it could be concluded that the current satellite network cost recovery process was a success.

However, he noted further, the paper satellite problem is actually only partially resolved. Despite the bureau's best recent efforts, in cooperation with the notifying administrations, to remove unused network registrations from the MIFR, the register currently contains registrations for many satellite networks that are no longer (or have never been) in use. These vestigial entries in the register have the effect of continuously blocking new entrants from accessing the spectrum/orbital resources tied to these registrations. Thus, "[m]ore effective means of preventing 'paper satellites' to remain in the MIFR "should be considered (ITU Doc. RAG11-1/1(Rev-1)-E, 2011, p. 2).

The current satellite network cost recovery process of Council Decision 482 only applies to the ITU's costs of processing, examining and publishing satellite network filings in a satellite project's initial stages of advance publication, coordination and notification. After the initial notification, there are no charges for maintaining the registration in the MIFR for the lifetime of the network, even if the satellite is replaced in the future by a spacecraft with similar characteristics. There is no recovery of the bureau's costs for maintaining the registration, which includes performing technical examinations to assess new networks' compatibility with existing ones recorded in the MIFR. The director concluded that "newcomers are taking on a greater financial burden than incumbents, many of which having been recorded prior to the entry of Decision 482. He continued:

> In order to ensure a more equitable apportionment of overall processing costs between satellite users, in particular with respect to the costs associated to the maintenance of frequency assignments during the lifetime of a satellite network, a fee model that would include yearly fees (related to the bureau's cost for the maintenance of information in the MIFR) might be considered [by Council and WRC]. (ITU Doc. RAG11-1/1(Rev-1)-E, 2011).

In view of the reluctance of ITU member states to reopen the huge debates of the past on the issue of satellite network cost recovery that had finally culminated in Decision 482, the RAG was not at all receptive to the director's ideas. In classic ITU fashion, it punted the issue to other bodies for further consideration. In its report to the director, the RAG observed:

> [T]he objectives of satellite cost recovery are neither to generate revenue for the Union nor to ensure the application of the principle of equitable access for the use of the spectrum/orbit resource; nor are the objectives to deal with satellite backlog and paper satellites. Consequently, in any further development of the issue, the … various arguments … especially the strong concerns expressed on any possible revision to Council Decision 482 (modified 2008), need to be considered by competent entities such as the Council and the Plenipotentiary Conference as regards the policy and financial aspects and a competent WRC as regards the regulatory aspects. (ITU Administrative Circular CA/199, 2011, p .3)

*A few months after the RAG,* the 2011 session of the Council was convened, which provided an opportunity for further consideration of the possible expansion of satellite network cost recovery. Costa Rica submitted a contribution to the Council supporting the director's concept of expansion. After noting Article 44's exhortation that radio frequencies and the associated orbital positions are limited natural resources that must be used efficiently to ensure that countries have equitable access to those resources, it described the difficulties experienced by countries in obtaining access to support their new satellite projects. Indeed, it stated, we "are faced with the apparent saturation of the orbit/spectrum resource." (ITU Doc. C11/73-E, 2011, p. 2) Repeating the director's assertions to the RAG, Costa Rica concluded that "in order to guarantee the equitable distribution of total processing costs between satellite users," Decision 482 should be revised, despite the fact that "it is very unlikely that cost recovery alone will be enough to avoid saturation." (ITU Doc. C11/73-E, 2011) Costa Rica urged the Council to form a working group to extend the satellite network cost recovery process to the costs of maintaining the MIFR and, to "[i]nvestigate the possibility of setting annual quotas for orbit/spectrum use by satellite networks, with the aim of promoting efficient and rational use of this resource." (ITU Doc. C11/73-E, 2011)

Russia also submitted a contribution to the Council discussing the director's report to the RAG on the extension of satellite recovery, but Russia opposed the idea. Russia agreed that the "paper satellite" problem continues to exist, but questioned whether charging for maintaining MIFR entries would be an effective solution and whether it would ultimately be equitable. Russia noted further that cost recovery payments for maintaining registrations for a well-heeled commercial mega-fleet provider might be more easily borne than the costs to a developing country for a single network providing vital national services. Moreover, Russia expressed concerns with undermining the meaning of a registration in the MIFR and how such fees would be computed. The better approach would be to address difficulties with the current Radio Regulations, including:

- Modification of Resolution 49 to introduce new criteria for confirming bringing into use, and regular updating of information to reflect changes. Introduction of a unique identifier applicable to a satellite as a physical object, etc.
- Monitoring of the satellite orbit and emissions.
- Refinement of the definition of regular operation (Nos. 11.44 and 11.47 of the Radio Regulations).
- Definition of the minimum period of use of a frequency assignment in order to qualify for bringing into use (at least three months), etc. (ITU Doc. C/11-75-E, 2011, p. 3)

As one might imagine, these contributions inspired considerable discussion. The majority of councilors agreed that: "the membership is facing difficulties with regard to equitable access to the orbit/spectrum resource due to various complex regulatory issues … which are within the purview and mandate of WRC" (ITU Doc. C11/89(Rev.1)-E., 2011, pp. 44–45). The ITU Council concluded that: "It is imperative to retain the current version of Decision 482 MOD 2008 without any changes unless there are in-depth studies carried out which could recommend the need to revise that Decision based on sound and valid arguments/ justification." (ITU Doc. C11/89(Rev.1)-E., 2011, p. 45) Thus, the director was instructed to bring the matter to the attention of WRC-12 and report back to Council-12 for any further action.

As instructed by the Council, the director submitted to WRC-12 a report summarizing the Council's discussions on whether to expand satellite network cost recovery to include maintenance of a registration in the MIFR. However, there were no proposals from member states on this topic. Only Costa Rica intervened to promote the extension of satellite cost recovery to include annual fees for maintaining recorded assignments in the MIFR. Costa Rica discussed the social good of satellite networks and the importance of access to the spectrum/orbital resource. The representative said that Costa Rica planned to bring a contribution to the next session of the Council in furtherance of this concept.

Iran, which is not a member of the Council, supported Costa Rica, noting that the issue raises the concern of developing countries' ability to obtain equitable access. The conference took no action on satellite network cost recovery. However, it did make many amendments to the Radio Regulations under Agenda Item 7 in an effort to improve the overall situation of apparent congestion of the spectrum/orbital resources and the inability of countries to gain access to these resources.

Following the conference, the director later reported to the 2012 session of the Council that the WRC-12 had achieved "a breakthrough" with respect to improvements to the Radio Regulations to limit the activities that contribute to congestion and misuse of ITU processes:

- By establishing a minimum continuous period of three months of operation of a GSO space station with transmitting and receiving capability deployed and maintained at the notified orbital position to consider that a satellite network has been brought into service (Nos. 11.44, 11.44.1, 11.44.2, 11.44B of the Radio Regulations).
- By clarifying administrations, Radiocommunication Bureau and Radio Regulations Board's actions in the application of No. 13.6 of the RR, in particular administrations' obligation to provide information on the actual use of the notified characteristics of commercial satellite networks if requested by the bureau.
- By identifying the formal list of specific satellite networks with which coordination needs to be effected (Nos. 9.36.2, 9.41, 9.42, 9,42.1 of the RR).
- By requesting the Radiocommunication Bureau to enquire on situations where the same satellite may have been used to maintain the rights of inactive satellite networks at various orbital locations by "hopping" from one location to another.
- By confirming previous decisions by the Radio Regulations Board of suppressing the rights of networks which had been maintained using the above approach. (C12/38-E, 2012, p. 2)

## WRC-12 Actions on Equitable Access (Resolution 80)

The continuing quest for interpretation and practical application of the concept of equitable access per Article 44 of the ITU Constitution also continued at WRC-12 Action on Resolution 80 (Rev. WRC-07): "Due diligence in applying the principles embodied in the Constitution," is a standing agenda item for every WRC's consideration (Agenda Item 8.1.3). The subject also arose in numerous other contexts throughout the conference. Many of the actions taken by WRC-12 advanced, directly or indirectly, the principle of equitable access.

The Radio Regulations Board (RRB) provided WRC-12 with a report on Resolution 80, as instructed by the terms of that Resolution, to review and make recommendations on linking the formal notification, coordination and registration procedures with the principles contained in Article 44 of the Constitution and No. **0.3** of the Preamble to the Radio Regulations, and to report to each future World Radiocommunication Conference with regard to this resolution." (ITU Radio Regulations, 2008, Volume III-E, p. 84)

The board provided advice on several areas of potential improvement to the Radio Regulations to alleviate the problems of congestion that frustrate entrants into GSO. Many of the board's suggestions were actively considered alongside the more than 600 proposals to the conference submitted under Agenda Item 7.

Several of Iran's proposals to the WRC-12 evoked Article 44. One sweeping proposal requested the conference to adopt an agenda item for the next WRC to perform a general overhaul of all of the regulatory provisions governing the use of space services. In support, Iran observed that the current regulatory system is fifty years old; it has been amended in piecemeal fashion; plus, there are "unbelievable" numbers of filings and "non-realistic" periods of validity for assignments in the MIFR. With all this in mind, there is a relative disadvantage for developing countries in coordination negotiations. Iran asserted that a new approach should be developed in time for the next World Radiocommunication Conference in 2015, which would "guarantee equitable access to all countries, in particular developing countries and those other countries which have either less used or not yet used these resources … in a most efficient, most effective and most economic manner." (ITU Doc. CMR12/75 (Add.11)-E, 2012, p. 6)

In the ensuing discussions at the conference, Iran backed down from its grand proposal but still insisted in the alternative on broadening and strengthening the language of standing Agenda Item 7 to allow for such an overhaul in the future, and to add a direct reference to Article 44 to the agenda item. After much discussion, a compromise solution was forged to modify the standing agenda item as follows (additions to text underscored): to consider possible changes, *and other options*, in response to Resolution 86 (Rev. Marrakesh, 2002) of the Plenipotentiary Conference, an advance publication, coordination, notification and recording procedures for frequency assignments pertaining to satellite networks, in accordance with Resolution **86 (Rev.WRC-07)** *to facilitate rational, efficient, and economical use of radio frequencies and any associated orbits, including the geostationary-satellite orbit* (Final Acts of WRC-12, Resolution 807 (WRC-12), 2012, p. 5).

Mexico and Bulgaria were the only administrations to formally submit proposals under Resolution 80, Agenda Item 8.1.3. Mexico proposed to revise Resolution 2 (Rev. WRC-03), "Equitable use, by all countries, with equal rights, of the geostationary-satellite and other satellite orbits and of frequency bands for space radiocommunication services" to provide priority access for national satellite systems for national security, public protection, or social welfare, relative to filings of other administrations whose territory is not within the coverage area of the proposed space system (ITU Doc. CMR12/18(Add.1)-E, 2012, pp. 2–3). Mexico later withdrew this proposal in favor of separate coordination discussions with a major commercial satellite operator. Bulgaria's proposal was narrowly tailored, seeking an exception to provisions of Appendix 30A Broadcasting-Satellite Service plan to enable it to complete its network (ITU Doc. CMR12/58(Add.3)-E, 2012).

The issue of equitable access was also a theme that arose under the conference's consideration of WRC-12 Agenda Item 1.13 on the treatment of the proliferation of satellite network filings for a new broadcasting-satellite service allocation in the 21.4–22 GHz band in ITU Radio Regions 1 and 3 (which includes all but the Western Hemisphere) to enable high resolution high definition television (HDTV) systems in those geographic regions. By December 2011, there were 890 advance publication filings and 291 coordination requests pending before the bureau for this particular allocation (ITU Doc. CMR12/4(Add.7) (Add.2)-E, 2012, p. 2). One of the fundamental questions facing the conference was how to allow administrations access to this band, including newcomers, in light of the tremendous interest. Previous conferences had made the decision to retain this as a "first come, first served" allocation, and not make it part of a plan:

> That a priori planning is not necessary and should be avoided as it freezes access according to technological assumptions at the time of planning and then prevents flexible use taking into account real world demand and technical developments. (Resolution 551 (WRC-07), ITU Radio Regulations, 2008, Vol. III, p. 289)

Instead, a unique regulatory and procedural approach was developed and agreed for this allocation in order to promote equitable access and to prevent paper satellites. Some worry, however, that this approach sets a dangerous precedent for departing from the first-come, first-served process in the non-planned bands.

Under this procedure, administrations that had submitted filings in the band were required to be:

> in compliance with Article 44 of the Constitution, review their submissions in the band 21.4-22 GHz submitted before 18 February 2012, with a view to reducing the number of their submissions to the absolute minimum necessary, and indicate to the Bureau before 30 June 2012, the networks which are no longer required to be considered and processed by the Bureau and administrations under Articles **9** and **11.** (Resolution 555 (WRC-12), ITU Radio Regulations, 2012, Vol. III-3, p. 307)

In addition, Resolution 555 (WRC-12) – "Additional regulatory provisions for broadcasting-satellite service networks in the band 21.4–22 GHz in Regions 1 and 3 for the enhancement of equitable access to this band" – urged administrations "to make the utmost efforts to accommodate submissions received from other administrations with few filings, especially covering their own territories." The resolution further invited the Council to exempt from satellite network cost recovery and to modify Decision 482 for networks refiled under this resolution's provisions.

Second, the conference introduced a controversial new procedural device to provide priority treatment for a single filing from those administrations with fewer filings. Resolution 553 (WRC-12), "Additional regulatory measures for broadcasting-satellite service networks in the band 21.4–22 GHz in Regions 1 and 3 for the enhancement of equitable access to this band," sets up a method for "queue jumping" under which, under defined circumstances, an administration may move a single satellite network filing ahead in the processing queue, thus jumping ahead in terms of priority over earlier filings.

Lastly, the conference adopted a special administrative due diligence procedure for satellite networks filings in this band (including those already submitted but not brought into use). The procedure calls for far more information than does Resolution 49, including something called an "ITU ID number," which Russia and the United States opposed in the wee hours of an all-night plenary session in the conference's final days. Moreover, the due diligence information must be supplemented following suspension, replacement of spacecraft, or end of life. A transitional measure requires such due diligence filings for networks already brought into use. These procedures are contained in Resolution 552 (WRC-12), "Long term access to and development in the 21.4–22 GHz band." Time will tell whether these unique regulatory mechanisms will be applied more broadly in the future to address issues of over-filing, apparent congestion, and equitable access.

Finally, Article 44 was the basis of new Resolution 11 (WRC-12), "Use of satellite orbital positions and associated frequency spectrum to deliver international public telecommunication services in developing countries." This resolution developed from consideration of a proposal by the members of the African Telecommunications Union seeking to preserve the "common heritage" status of the orbital positions recorded by Intelsat when it was an international satellite organization prior to its privatization. The African administrations sought to ensure that these orbital resources would be permanently maintained for global satellite connectivity for delivery of international public telecommunication services at an affordable price to developing countries. This topic had been brought to the attention of the Radiocommunication Bureau in 2009 by Colombia and was the subject of proposals at the 2010 Plenipotentiary Conference by African nations. Although the existing regulatory procedures did not contemplate such action, WRC-12 resolved to collaborate with the ITU development sector on satellite technologies, applications, and regulatory procedures to assist developing countries to implement satellite networks and services. Further, it was resolved that the radiocommunication sector would "undertake studies to determine whether it might be necessary to apply additional regulatory measures to enhance the availability of public international telecommunication services delivered through satellite technology." (Resolution 11 (WRC-12), ITU Radio Regulations, 2012, Vol. III-E, p. 17) Separately, the United States, as Intelsat's notifying administration, agreed to a special designation in the MIFR for Intelsat's original assignments.

## Council Postscript

The annual meeting of the Council following WRC-12 noted the current situation concerning satellite network cost recovery (Table 7.2).

**Table 7.2** Status of the implementation of Decision 482 for 2010–2011 (ITU Doc. C12/22-E, 2012, p. 2)

| | | 2010 | 2011 |
|---|---|---|---|
| Total invoices issued[a] (including free entitlement) | CHF | 13 266 544 | 14 374 442 |
| Free entitlement | CHF | 1 079 528 | 1 097 377 |
| Payments received | CHF | 11 908 225 | 12 218 430 |
| Percentage of invoices issued in 2010/2011, due by 31.12.2011 and having been paid | | | 99 % |

[a]99 % of the invoices were paid in a timely manner. Invoices are payable within 6 months of the date of issue

The ITU Council also received the report from WRC-12's discussions on satellite cost recovery. With a familiar ring, the report observed that the amendments to the Radio Regulations to improve the situation of congestion and over-filing would not come into force until January 2013. Thus, additional time would be needed before the impact of these changes could be assessed. The Council was therefore invited to defer the topic of expansion of cost recovery to future sessions (ITU Doc. C12/38-E, 2012, p. 2).

ITU Council-12 did make a small modification to update Decision 482 consequential to WRC-12's action under Resolution 555 to exempt from cost recovery requirements certain broadcasting-satellite service network filings in the band 21.4–22 GHz. Thus, Decision 482 was modified by Council-12, to add this exclusion and also to remove a provision requiring an external biennial audit. The cost recovery methodology of Council 2005 remains unchanged.

Again in 2013, the Secretary-General reported a 99 % payment rate for satellite network cost recovery notices, with invoices totaling more than CHF 14.5 million. Some further minor modifications were made to Decision 482 to make adjustments for the special case of filings in the band 21.4–22 GHz per the decisions of WRC-12. No proposals were made to expand satellite network cost recovery. The current version of Decision 482 is provided in Annex C.

## References

African Telecommunications Union Administrations: African Common Proposals for the Work of the Conference, Geneva. ITU Doc. CMR12/17(Addendum 29)-E (2012)

Arab Spectrum Management Group: Proposals for the Work of the Conference, Geneva. ITU Doc. CMR12/25(Addendum 28) (Addendum 1)-E (2012)

Bulgaria: Proposals for the Work of the Conference Agenda Item 8.1.3. Geneva. ITU Doc. CMR12/58(Addendum 3)-E (2012)

Canada: Proposals for the Work of the Conference: Agenda Item 7, Geneva. ITU Doc. CMR12/7-E(Addendum 4) (2012)

Chairman of the Standing Committee on Administration and Management. Report by the Chairman of the Standing Committee on Administration and Management. Geneva. ITU Doc. C11/89(Rev.1)-E (2011).

Costa Rica: Revising Charges for Satellite Network Filings, Geneva. ITU Doc. C/11/73-E (2011)
Department of State: United States Delegation Report World Radiocommunication Conference 2012. Washington. http://www.state.gov/documents/organization/108955.pdf (2012)
Director, Radiocommunication Bureau: Survey of Satellite Networks with BSS Frequency Assignments for Region 1 or 3 in the 21.4-22 GHz Band, Geneva. ITU Doc. CMR12/4 (Addendum 7) (Addendum 2)-E (2012).
Director, Radiocommunication Bureau: Survey of Satellite Networks with BSS Frequency Assignments for Region 1 or 3 in the 21.4-22 GHz Band, Geneva. ITU Doc. CMR12/4 (Addendum 7) (Addendum 2)-E (2012)
Director, Radiocommunication Bureau: Report of the Director on the Activities of the Radiocommunication Sector, Part 2: Experience in the application of the radio regulatory procedures and other related matters, Geneva. ITU Doc. CMR12/4(Corrigendum 1) (Addendum 2) (Revision 1)-E (2012)
Director, Radiocommunication Bureau: Report of the Director on the Activities of the Radiocommunication Sector, Part 1: Activities of the Radiocommunication Sector in the Period between WRC-07 and WRC-12, Geneva. ITU Doc. CMR12/4(Addendum 1) (Revision 1)-E (2012)
Director, Radiocommunication Bureau: Report to the Eighteenth Meeting of the Radiocommunication Advisory Group, Geneva. ITU Doc. RAG11-1/1 (Rev-1)-E (2011)
European Conference of Postal and Telecommunications Administrations: European Common Proposals for the Conference: Issues Related to Clarification of Bringing into Use and Due Diligence of Satellite Networks, Geneva. ITU Doc. CMR12/5(Addendum 28) (Addendum1)-E (2012)
Federal Communications Commission. Public Notice. FCC Seeks Comment on Recommendations Approved by the Advisory Committee for the 2012 World Radiocommunication Conference. Washington. http://hraunfoss.fcc.gov/edocs_public/attachmatch/DA-11-712A1.pdf (2011). Accessed 20 April
France: WRC-12 Agenda Item 7—Proposed Draft of CPM Text on Bringing into Use of Assignments to Geostationary Satellite Networks in Certain Radio Services and Frequency Bands, Geneva. ITU Doc. SC/43-E (2010).
International Telecommunication Union: ITU Radiocommunication Bureau 2011 Annual Space Report to the STS-12 on the use of geostationary-satellite orbit and other orbits, Geneva, 4 January 2011. http://www.itu.int/ITU-R/space/snl/report/index.html (2012)
International Telecommunication Union: Final Acts of the World Radiocommunication Conference (Geneva, 2012). International Telecommunication Union, Geneva (2012)
International Telecommunication Union: Decision 482 (Modified 2012), Implementation of Cost Recovery for Satellite Network Filings, Geneva. ITU Doc. C12/95(Rev.2)-E (2012)
International Telecommunication Union: Administrative Circular. CA/199. International Telecommunication Union, Geneva (2011)
International Telecommunication Union: Radio Regulations. International Telecommunication Union, Geneva (2008)
Iran: Proposals for the Work of the Conference Agenda Item 8.2, Geneva. ITU Doc. CMR12/75(Addendum 11)-E (2012)
Mexico: Proposals for the Work of the Conference Agenda Item 8.1.3, Geneva. ITU Doc. CMR12/18(Addendum 12)-E (2012)
Russia: Introduction of an Annual Fee for Maintaining a Frequency Assignment to a Satellite Network in the Master International Frequency Register (MIFR), Geneva. ITU Doc. C11/75-E (2011)
Secretary-General: Summary of the Discussion on Cost Recovery for the Processing of Satellite Network Filings, Geneva. ITU Doc. C12/38-E (2012)
Secretary-General: Cost Recovery for the Processing of Satellite Network Filings, Geneva. ITU Doc. C12/22-E (2012)

Secretary-General: CPM Report on Technical, Operational and Regulatory/Procedural Matters to be considered by the 2012 World Radiocommunication Conference, Geneva. ITU Doc. CMR12/3-E (2012)

Secretary-General: Agenda of the Conference. Geneva. ITU Doc. CMR12/1-E (2012)

Special Committee on Regulatory/Procedural Matters: Report of the Special Committee to the Conference Preparatory Meeting, Geneva. ITU Doc. CPM-2/2E (2011)

United Kingdom: WRC-12 Agenda Item 7—Bringing into Use Assignments to Satellite Networks, Geneva. ITU Doc. SC/33-E (2010)

United States of America: Modifications to Resolution 49 (Rev.WRC-97), Geneva. ITU Doc. CPM11-2/150-E (2011).

# Chapter 8
# 2015 and Beyond

## Preparations for WRC-15

At the time of this writing, preparations are well under way for the next World Radiocommunication Conference, which will convene in Geneva in November 2015. The review of the satellite regulatory provisions is again proving to be a major theme of the conference preparations. Other key issues involve the protection of satellite spectrum from reallocation to other services as well as obtaining additional allocations for fixed and mobile-satellite services (Fig. 8.1).

As indicated at WRC-12, Iran has proposed during preparatory activities for the 2015 conference that a comprehensive review, or overhaul, of the regulatory regime governing space services in non-planned frequency bands be undertaken. It has proposed to ITU-R Working Party 4A and the Working Party of the Special Committee on Regulatory/Procedural Matters that such a review take place under Agenda Item 7:

> The issue of [the] radio regulatory regime governing space services has been the subject of discussions and debate at many WARCs and WRCs since its creation at [E]ARC-63. Relevant articles of the former ITU Convention and current ITU Constitution provide necessary principles to be observed in using orbital/spectrum resources for space services in an equitable, efficient and economical use, taking into account the particular situation of developing countries.
>
> However, [the] current situation caused serious difficulties which are being encountered by many countries, in particular, those of developing ones to have timely equitable access to orbital/spectrum resources. These difficulties have been reported to the ITU Council and several other ITU entities such as CPM-12 under WRC-15 Agenda item 7 (ITU Doc. SC-WP/11-E 2013, p. 1).

Iran points to a long list of examples of the sources of these difficulties, including:

- excessive filings (warehousing of the orbital/spectrum resources) in the form of submission of multiple advance publication and multiple request[s] for coordination at every three degrees spacing of orbital positions …



A.L. Allison, *The ITU and Managing Satellite Orbital and Spectrum Resources in the 21st Century*, SpringerBriefs in Space Development, DOI 10.1007/978-3-319-05314-1_8, 

**Fig. 8.1** ITU flags adorn the Pont du Mont Blanc in Geneva to mark the occasion of the World Radiocommunication Conference in January 2012 (ITU File Photo)

- excessive and/or misuse of application of certain provisions of the Radio Regulations such as RR No. **11.41** … [and] Ambiguity and/or lack of clarity and deficiencies in the meaning and scope of application of certain other provisions such as RR No. **11.44** …
- submission of technical characteristics designed to gain over protection and to theoretically reduce the emitted power towards other administration's territory, and/or satellite services …
- total mismatch between the satellites that are in actual use and those claimed to be in operation in such a manner that one single satellite theoretically protects multiple orbital positions …
- catastrophic status of Master Register, which contains almost 50 % of no coordinated assignments… (ITU Doc. SC-WP/11-E 2013, pp. 1–2).

Based on this review, Iran proposes "to examine the entire regulatory regime governing the use of the orbital/spectrum resources in the space services in non-planned bands as contained in RR Articles **9, 11, 13, 14**, and **15** and their associated Rules of Procedure as well as all relevant appendices and annexes and resolutions of previous WRCs with a view to give a new and fresh look for comprehensive review of applicable regulatory regime to space services in non-planned bands" (ITU Doc. SC-WP/11-E 2013, p. 4).

Other participants in the preparations, including the United States, Russia and western European nations, have opposed such a wholesale review, instead opting for continuation of the current incremental process for reviewing and improving the

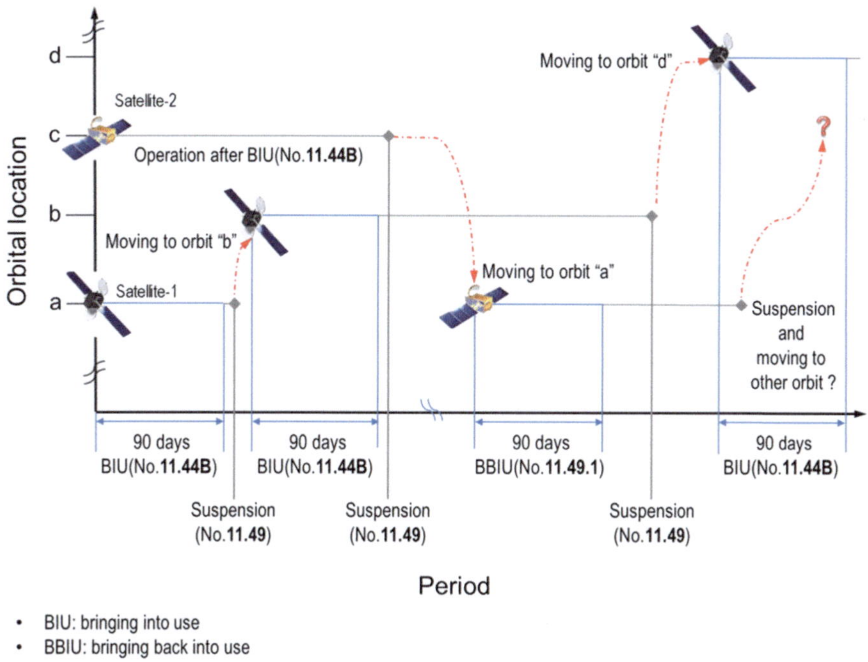


**Fig. 8.2** Korean contribution to the working party of the Special Committee on Regulatory/Procedural Matters, illustrating an example of possible misuses of the provisions of Nos. 11.44B and 11.49 using two satellites (ITU Doc. SC-WP/21-E. 2013, p. 3)

space regulatory procedures provided for under Agenda Item 7. The United States responded to Iran's proposal:

> While acknowledging that the underlying principle of [Iran's proposal] is [that] the current methodology for registering a satellite network can be sometimes cumbersome and has "loopholes" that allow some administrations to exploit the process, our overall assessment is that the current satellite registrations process is working. The current process provides a stable pathway for administrations to meet the requirements for satellite registration. Looking at all the modifications made since WRC-2000 to the advanced publication, coordination and notification processes, it is clear that these modifications take into account principles of equitable, efficient, and economical use of the scarce orbital and spectrum resources. Throughout these changes, the satellite registration process has remained relatively stable. It is critical that any proposed general overhaul or wholesale changes to the satellite registration process preserves the rights and satisfy the needs of currently registered satellite networks and balances these with the rights and needs of new satellite networks seeking entry in the ITU Master International Frequency Register (MIFR) (ITU Doc. 4A/286-E 2013, p. 1) (Fig. 8.2).

Many such incremental improvement proposals are already being prepared for consideration under Agenda Item 7 at WRC-15. These proposals include: (i) eliminating the advance publication stage of the process; (ii) clarifying the language of the Radio Regulations regarding bringing into use and suspension; (iii) proposing new measures to prevent misuse of the processes via satellite drifting and leasing; and (iv) potentially changing the basis for determination of when coordination is necessary.

Meanwhile, Uganda and Tanzania have proposed a course of studies under Agenda Item 9.1.3 (equitable access) pursuant to WRC-12 Resolution 11 "to determine whether it might be necessary to apply additional regulatory measures to enhance the availability of the public international telecommunications delivered through satellite technology" (ITU Doc. 4A/264-E 2013, p. 1).

## Conclusion and Postscript

This book has elaborated on the ongoing efforts of the International Telecommunication Union and its members to manage a globally shared natural resource that is in high demand by nations and companies and is also subject to treaty mandates of equitable access and promoting the advancement of technology, efficient operation, and rapid implementation of services.

As the number of filings for new satellite networks increased beyond the realm of real needs, and the ITU's ability to process them, the ITU community was forced to find solutions—or to consider new alternatives. They decided that the ITU remains the best forum for managing the orbital/spectrum resource and that the processes and structures in effect can benefit from ongoing incremental improvement. This incremental approach was favored over wholesale re-creation or replacement of the current system, which could have the effect of destabilizing the industry. Such a result could imperil existing infrastructure and future investments needed for research and development that could further advance technologies and services. The ITU's adoption of administrative due diligence, combined with discrete regulatory reforms from one conference to the next, has proven to be a reasonable and measured way of building acceptable solutions over time in the cadence of the development of new technology, services, and growing demand.

Similarly, with the adoption of satellite network cost recovery, the union has evolved its processes as the composition of its membership has grown. As the global telecommunication environment has transformed over time, it has had the unintended consequence of eroding the union's financial base due to the need for more services but without a commensurate increase in revenues. After many years of painful trial and error, the ITU has found an acceptable way to recover the costs of services provided to consumers of satellite network processing services while accommodating the special needs of developing countries. It also yielded the additional benefit of helping to lessen the number of virtual satellite filings as had been the original goal of the proposed financial due diligence method.

Equitable access is not a fixed point to be achieved. Rather it is a goal that the ITU and its members must continue to strive for as technology and the external environment continue to change—as do the needs of ITU members. Consideration of Resolution 80, the standing WRC agenda items, and Article 44 of the ITU constitution serve to keep the goal of equitable access always in view and under consideration. These principles of equitable access are ever relevant to the union's continual adaptation to change in order benefit its members and the world's citizens in relation to satellite services. The ITU's membership-driven work and consensus-based decision making naturally leads to serious consideration of these difficult issues. The hope is that this leads to the development of solutions over time that are acceptable to all parties.

For nearly 150 years, the ITU, with its unique blend of government and private industry members, has managed, through good will, cooperation, mutual desire, and perseverance, to continuously adapt as necessary to benefit from ongoing technological advances and to respond to changes in the larger environment. The ITU and its members have created a durable institution that is capable of meeting the evolving needs of its members. In this way, the ITU's management of access to geostationary satellite orbit will continue to meet the needs of the world's community into the future, supporting a robust industry and vital national services.

However, it must be observed that the union's long and historic success has recently been called into question as a consequence of its recent conference to revise its other administrative regulations, the International Telecommunication Regulations (ITRs). The 2012 World Conference on International Telecommunications (WCIT-12) in Dubai was the first effort by the union to update the ITRs since 1988. The ITRs traditionally address matters concerning global interconnection and interoperability of telecommunication networks, such as telegraphy (originally) and then telephony. The issue before the Dubai conference included debate over the ITU's jurisdiction and appropriate role with respect to the Internet, including issues related to cyber security and spam. Many developing countries, including the Arab nations, China and Russia, proposed a greater role for the ITU in the management of the Internet. The conference was unable to achieve the usual consensus, and it concluded with a vote. Fifty-four countries, including the United States, the United Kingdom, Canada, Australia, and most of western Europe, refused to sign the Final Acts, mainly due to concerns with preservation of Internet freedom (ITU Final Acts of the World Conference on International Telecommunications, Dubai 2012 a, b). The unresolved issues of Dubai will undoubtedly arise again at the ITU's next Plenipotentiary Conference in Busan, Korea, in October–November 2014.

For a century and a half, the ITU has evolved on a parallel track with the development of telecommunications technologies and services, from telegraphy to telephony; from wireless telegraphy to broadcasting to space communications. The role of the ITU with respect to the Internet that the world eventually agrees upon will be the next chapter in its long history. That notwithstanding, the significance of its role over the management of orbital/spectrum resources for space communications remains vital and unquestioned. The ITU will remain the manager of the world's resources for satellite communications for the twenty-first century.

## References

International Telecommunication Union: Final Acts of the World Conference on International Telecommunications, Dubai. International Telecommunication Union, Geneva (2012a)

International Telecommunication Union: Final Acts of the World Radiocommunication Conference, Geneva. International Telecommunication Union, Geneva (2012b)

Iran: Possible Option on Comprehensive Review of the Radio Regulatory Regime Governing Space Services in Non-planned Bands, Geneva. ITU Doc. SC-WP/11-E (2013)
Korea: Review of the Provisions of Nos. **11.44B** and **11.49** of the Radio Regulations, Geneva. ITU Doc. SC-WP/11-E (2013)
Uganda, Tanzania: Preliminary reflection with regard to the studies under Resolution **11 (WRC-12)**: Use of the orbital slots and associated frequency spectrum to deliver international public telecommunication services, Geneva. ITU Doc. 4A/264-E (2013)
United States of America: Further Thoughts on Simplification of Radio Regulation Regime on Space Services, Geneva. ITU Doc. 4A/286-E (2013)
World Radiocommunication Conference: (WRC-15) Agenda and Relevant Resolutions: http://www.itu.int/dms_pub/itu-r/oth/12/01/R12010000014A01PDFE.pdf (2013)
ITU Website for World Radiocommunication Conference : http://www.itu.int/en/ITU-R/conferences/wrc/2015/Pages/default.aspx (2015)
ITU Website on WCIT-12: Conference Overview: http://www.itu.int/en/wcit-12/Pages/overview.aspx

# Appendix A: Resolution 18 (1994) Review of the ITU's Frequency Coordination and Planning Framework for Satellite Networks

The Plenipotentiary Conference of the International Telecommunication Union (Kyoto, 1994),

*Considering*

(a) That Article 44 of the Constitution (Geneva, 1992) sets out the basic principles for the use of the radio-frequency spectrum and the geostationary-satellite orbit.
(b) The increasing globalization and diversification of telecommunication systems, particularly satellite networks.
(c) That there is growing concern about the accommodation of new satellite networks, including those of new ITU Members, and the need to maintain the integrity of ITU procedures and agreements.
(d) That the report of the Voluntary Group of Experts on Simplification of the Radio Regulations to be considered at the 1995 World Radiocommunication Conference (WRC-95) maintains the current coordination procedures, albeit in simplified form.
(e) That the agenda for WRC-95, and the provisional agenda for the 1997 World Radiocommunication Conference (WRC-97), includes consideration of the broadcasting-satellite plans for Regions 1 and 3 in Appendices 30 and 30A of the Radio Regulations.
(f) That radiocommunication study groups are considering possible improvements to these plans, taking into account that since the plans were devised more modern technologies, including digital techniques, have emerged which may provide more effective and affordable alternative options for the provision of services.
(g) That the radiocommunication study groups are also developing technical coordination procedures for satellite networks and have requested the Regulatory Working Party of the Conference Preparatory Meeting (1995) to develop complementary regulatory provisions.
(h) The concerns of some Members about lack of compliance with coordination procedures.
(i) That many developing countries are in need of assistance in the implementation of satellite network coordination procedures.

A.L. Allison, *The ITU and Managing Satellite Orbital and Spectrum Resources in the 21st Century*, SpringerBriefs in Space Development, DOI 10.1007/978-3-319-05314-1, 

*Resolves to instruct the Director of the Radiocommunication Bureau*

1. In consultation with the Radiocommunication Advisory Group and taking into account inputs from the Radio Regulations Board (RRB), to initiate a review of some important issues concerning international satellite network coordination, including:

   (a) Linkages between ITU procedures and commitments to take up notified frequencies and orbital positions.
   (b) The ongoing need for the ITU's frequency coordination and planning framework for satellite networks to continue to be relevant to rapidly advancing technological possibilities in order, for example, to facilitate the establishment of multiservice satellite systems.
   With the objectives of:

   i. Ensuring equitable access to the radio-frequency spectrum and the geostationary-satellite orbit, and the efficient establishment and development of satellite networks.
   ii. Ensuring that international coordination procedures meet the needs of all administrations in establishing their satellite networks, while at the same time safeguarding the interests of other radio services.
   iii. Examining technological advances in relation to the allotment plans with the aim of determining whether they foster the flexible and efficient use of the radio-frequency spectrum and the geostationary-satellite orbit.

2. To ensure that this review takes account of the ongoing work of the Radiocommunication Sector and, in particular, in the RRB and in radiocommunication study groups.
3. To coordinate activities, as necessary, with the Directors of the other two Bureaux.
4. To make a preliminary report to WRC-95 and a final report to WRC-97.

*Instructs the Secretary-General*

To encourage the participation of all interested parties, including satellite system operators, at an appropriately high level, and to afford the Director all necessary assistance for the successful completion of the review.

# Appendix B: Resolution 80 (Rev.WRC-07) Due Diligence in Applying the Principles Embodied in the Constitution

The World Radiocommunication Conference (Geneva 2007),

*Considering*

(a) That Articles 12 and 44 of the Constitution lay down the basic principles for the use of the radio-frequency spectrum and the geostationary-satellite and other satellite orbits.
(b) That those principles have been included in the Radio Regulations.
(c) That Article I of the Agreement between the United Nations and the International Telecommunication Union provides that "the United Nations recognizes the International Telecommunication Union (hereinafter called "the Union") as the specialized agency responsible for taking such action as may be appropriate under its basic instrument for the accomplishment of the purposes set forth therein."
(d) That, in accordance with Nos. **11.30, 11.31** and **11.31.2**, notices shall be examined with respect to the provisions of the Radio Regulations, including the provision relating to the basic principles, appropriate rules of procedure being developed for the purpose.
(e) That WRC-97 instructed the Radio Regulations Board (RRB) to develop, within the framework of Nos. **11.30, 11.31** and **11.31.2**, rules of procedure to be followed in order to be in compliance with the principles in No. **0.3** of the Preamble to the Radio Regulations.
1. That the Board, in accordance with Resolution **80 (WRC-97)**, submitted a report to WRC-2000 suggesting possible solutions and stating that, after examining the Radio Regulations, it had concluded that there are no provisions currently in the Radio Regulations that link the formal notification or coordination procedures with the principles stated in No. **0.3** of the Preamble to the Radio Regulations.
(f) That the Legal Subcommittee of the Committee on the Peaceful Uses of Outer Space of the United Nations General Assembly has drawn up recommendations in this respect.

A.L. Allison, *The ITU and Managing Satellite Orbital and Spectrum Resources in the 21st Century*, SpringerBriefs in Space Development,
DOI 10.1007/978-3-319-05314-1, 

*Noting*

i. That, in accordance with the provisions of No. 127 of the Convention, the Conference may give instructions to the Sectors of the Union.
ii. That, according to No. 160C of the Convention, the Radiocommunication Advisory Group (RAG) shall review any matter as directed by a conference.
iii. The RRB report to WRC-2000 (see Annex 1).
iv. The RRB report to WRC-03 (see Annex 2).
v. That some of the issues identified in the report referred to in noting c) have been resolved before WRC-07.

*Resolves*

2. To instruct the Radiocommunication Sector, in accordance with No. 1 of Article 12 of the Constitution, to carry out studies on procedures for measurement and analysis of the application of the basic principles contained in Article 44 of the Constitution.
3. To instruct the RRB to consider and review possible draft recommendations and draft provisions linking the formal notification, coordination and registration procedures with the principles contained in Article 44 of the Constitution and No. **0.3** of the Preamble to the Radio Regulations, and to report to each future World Radiocommunication Conference with regard to this Resolution.
4. To instruct the Director of the Radiocommunication Bureau to submit to each future World Radiocommunication Conference a detailed progress report on the action taken on this Resolution.

*Invites*

1. The other organs of the Radiocommunication Sector, in particular the RAG, to make relevant contributions to the Director of the Radiocommunication Bureau for inclusion in his report to each future World Radiocommunication Conference.
2. Administrations to contribute to the studies referred to in *resolves* 1 and to the work of the RRB as detailed in *resolves* 2.

## Annex 1: To Resolution 80 (Rev.Wrc-07)

### *RRB Report to WRC-2000*

In the RRB Report to WRC-2000,[1] several members of the Board noted some difficulties likely to be experienced by administrations, particularly administrations of developing countries, as follows:

- The "first-come first-served" concept restricts and sometimes prevents access to and use of certain frequency bands and orbit positions.

[1] This Report can be found in Document 29 to WRC-2000.

- A relative disadvantage for developing countries in coordination negotiations due to various reasons such as a lack of resources and expertise.
- Perceived differences in consistency of application of the Radio Regulations.
- The submitting of "paper" satellites that restricts access options.
- The growing use of the bands of the Plans of Appendices **30** and **30A** by regional, multichannel systems, which may modify the main purpose of these Plans to provide equitable access to all countries.
- The considerable processing delays in the Radiocommunication Bureau are due to the very complex procedures required and the large number of filings submitted; these delays contribute to a coordination backlog of 18 months which could extend to 3 years and creates uncertain regulatory situations, additional delay in the coordination process that cannot be overcome by administrations, and the possible loss of the assignment because the allotted time is exceeded.
- Satellite systems may already be in orbit before completion of coordination.
- Statutory time-frames, such as those in No. **11.48**, may often be insufficient for developing countries to be able to complete the regulatory requirements as well as the design, construction and launch of satellite systems.
- No provisions for international monitoring to confirm the bringing into use of satellite networks (assignments and orbits).

## Annex 2: To Resolution 80 (Rev.Wrc-07)

### *RRB Report to WRC-03*

In the RRB Report to WRC-03,[2] concepts to satisfy *resolves* 2 of Resolution **80 (WRC-2000)** were provided, as follows:

Special measures for countries submitting their first satellite filing:

- On an exceptional basis, special consideration could be given to countries submitting their first filing for a satellite system, taking into account the special needs of developing countries.
- Such consideration should take into account the following:
- Impact on other administrations.
- Satellite service of the system (i.e. FSS, MSS, BSS).
- Frequency band covered by the filing.
- System is intended to meet the direct needs of the country(s) concerned.

Extension of the regulatory time-limit for bringing into use:

- Conditions could be specified under which extensions might be granted on an exceptional basis to developing countries when they are not able to complete the

---

[2] This report can be found in Addendum 5 to Document 4 of the WRC-03.

regulatory date requirements, so that sufficient time for design, construction and launch of satellite systems is made available.

- The conditions created under the previous paragraph should be included in the Radio Regulations as provisions that would allow the Radiocommunication Bureau to grant the extension.

# Appendix C: Decision 482 (Modified 2013) Implementation of Cost Recovery for Satellite Network Filings

The Council,

*Considering*

(a) A Resolution 88 (Rev. Marrakesh, 2002) of the Plenipotentiary Conference on the implementation of cost recovery for satellite network filings.
(b) Resolution 91 (Rev. Guadalajara, 2010) of the Plenipotentiary Conference on cost recovery for some products and services of ITU.
(c) Council Resolution 1113, on cost recovery for the processing by the Radiocommunication Bureau of space notifications.
(d) Document C99/68 reporting on the Council Working Group on implementation of cost recovery for satellite network filings.
(e) Document C99/47 on cost recovery for some ITU products and services.
(f) Document C05/29 on cost recovery for the processing of satellite network filings.
(g) That WRC-03 and WRC-07 adopted provisions referring to Council Decision 482, as amended, under which a satellite network filing is cancelled if payment is not received in accordance with the provisions of this decision.
(h) That WRC-07 significantly revised the regulatory procedures associated to the fixed-satellite service Plan contained in Appendix 30B that entered into force as of 17 November 2007.
(i) That the date of entry into force of Decision 482 (modified 2005) was 1 January 2006.

*Recognizing*

The practical experience of the Radiocommunication Bureau in implementing cost-recovery filing charges and the methodology as reported to the Council at its 2001 to 2007 sessions in accordance with Decision 482 as revised by the Council.

A.L. Allison, *The ITU and Managing Satellite Orbital and Spectrum Resources in the 21st Century*, SpringerBriefs in Space Development,
DOI 10.1007/978-3-319-05314-1, © Audrey Allison 2014

*Decides*

1. That all satellite network filings concerning advance publication, their associated requests for coordination or agreement (Article 9 of the Radio Regulations (RR), Article 7 of Appendices 30/30A to the RR, Resolution 539 (Rev.WRC-03)), the use of the guardbands (Article 2A to Appendices 30/30A to the RR), requests for modification of the space service plans and lists (Article 4 of Appendices 30 and 30A to the RR), requests for the implementation of the fixed-satellite service plan (former Sections IB and II of Article 6 of Appendix 30B to the RR up to 16 November 2007), and requests for the conversion of an allotment into an assignment with modification which is beyond the envelop characteristics of the initial allotment, the introduction of an additional system, modification of the characteristics of an assignment in the List of Appendix 30B to the RR (Article 6 of Appendix 30B to the RR as from 17 November 2007) shall be subject to cost-recovery charges if, and only if, they have been received by the Radiocommunication Bureau on or after 8 November 1998.

1bis That all satellite network filings concerning notification for recording of frequency assignments in the Master International Frequency Register (Article 11 of the RR, Article 5 of Appendices 30/30A to the RR and Article 8 of Appendix 30B to the RR) received by the Radiocommunication Bureau on or after 1 January 2006 shall be subject to cost-recovery charges if, and only if, they refer to advance publication or modification of the space service plans or lists (Part A), requests for the implementation of the fixed-satellite service plan or requests for the conversion of an allotment into an assignment with modification which is beyond the envelop characteristics of the initial allotment, the introduction of an additional system, the modification of the characteristics of an assignment in the List of Appendix 30B to the RR, as appropriate, received on or after 19 October 2002.

1ter That all requests for the implementation of the fixed-satellite service plan (former Sections IA and III of Article 6 of Appendix 30B to the RR) shall be subject to cost-recovery charges if, and only if, they have been received by the Radiocommunication Bureau on or after 1 January 2006.

1quater That all requests for consolidation of frequency assignments in the MIFR of different GSO networks submitted by an administration (or an administration acting on behalf of a group of named administrations) at the same orbital position into frequency assignments of a single satellite network received by the Radiocommunication Bureau on or after 1 July 2013, shall be subject to cost recovery charges,

2. That for each satellite network[3] filing communicated to the Radiocommunication Bureau, the following charges[4] shall apply:

---

[3] In this decision, the term "satellite network" refers to any space system in accordance with No. 1.110 of the Radio Regulations.

[4] The fee per "unit" (see Annex (Appendix C)) shall not be understood as a tax imposed on spectrum users. It is used here as a driver for the calculation of cost recovery relating to publication of satellite systems.

(a) For filings received up to and including 29 June 2001, Decision 482 (C-99) applies; these filings are charged at publication in accordance with the fee schedule in force at the date of publication.
(b) For filings received on or after 30 June 2001, but before 1 January 2002, Decision 482 (C-01) applies; these filings are charged at publication with a flat fee in accordance with the fee schedule in force at the date of receipt, and an additional fee (if any) according to the fee schedule in force at the date of publication.
(c) For filings received on or after 1 January 2002, but before 4 May 2002, Decision 482 (C-01) applies; the flat fee, calculated in accordance with the fee schedule in force at the date of receipt, is payable after receipt of the notice, and the additional fee (if any), calculated in accordance with the fee schedule in force at the date of publication, is payable after publication of the notice.
(d) For filings received on or after 4 May 2002, but before 31 December 2004, Decision 482 (C-02) applies; the flat fee, calculated in accordance with the fee schedule in force at the date of receipt, is payable after receipt of the notice, and the additional fee (if any), calculated in accordance with the fee schedule in force at the date of receipt, is payable after publication of the notice.
(e) For filings received on or after 31 December 2004 but before 1 January 2006, Decision 482 (C-04) applies; the flat fee, calculated in accordance with the fee schedule in force at the date of receipt, is payable after receipt of the notice, and the additional fee (if any), calculated in accordance with the fee schedule in force at the date of receipt, is payable after publication of the notice.
(f) For filings received on or after 1 January 2006 but before 1 January 2009 except those received under Appendix 30B as from 17 November 2007, Decision 482 (C-05) applies; the fee, calculated in accordance with the fee schedule in force at the date of receipt, is payable after receipt of the notice.
(g) For filings received on or after 1 January 2009, including those received under Appendix 30B as from 17 November 2007, but before 14 July 2012, Decision 482 (C-08) applies; the fee, calculated in accordance with the fee schedule in force at the date of receipt, is payable after receipt of the notice.
(h) For filings received on or after 14 July 2012, but before 1 July 2013, Decision 482 (C-12) applies; the fee, calculated in accordance with the fee schedule in force at the date of receipt, is payable after receipt of the notice.
(i) For filings received on or after 1 July 2013, Decision 482 (C-13) applies; the fee, calculated in accordance with the fee schedule in force at the date of receipt, is payable after receipt of the notice.

3. That the fee shall be regarded as a charge for a satellite network filing. There will be no charge for modifications which do not result in further technical or regulatory examination by the Radiocommunication Bureau, except modifications under 1quater above, including but not limited to the name of the satellite/earth station and its associated satellite name, name of the beam, responsible administration, operating agency, date of bringing into use, period of validity, associated satellite (and beam) or earth station name.

4. That each Member State shall be entitled to the publication of special sections or parts of the BR IFIC (space services) for one satellite network filing each year without the charges referred to above. Each Member State in its role as the notifying administration may determine which network shall benefit from the free entitlement.[5]
5. That the nomination of the free entitlement for the calendar year of receipt by the Bureau of the satellite network filing based on the formal date of receipt of the filing shall be made by the Member State no later than the end of the period for payment of the invoice in *decides* 9 below. The free entitlement cannot be applied to a filing previously cancelled for non-payment.
6. That for any satellite network for which the advance publication information (API) was received prior to 8 November 1998, there will be no cost-recovery charges for the first coordination request referring to that API, regardless of when the Radiocommunication Bureau receives it. Any modifications received on or after 1 January 2006 shall be subject to a charge in accordance with *decides* 2 above.
7. That there will be no cost-recovery charges for any Part A submission involving the application of Article 4 of Appendices 30/30A received by the Bureau prior to 8 November 1998 or Part B submission involving the application of Article 4 of Appendices 30/30A where the associated Part A was received prior to 8 November 1998. Any request for publication in Part A received after 7 November 1998 under §4.3.5 up to 2 June 2000 and then §4.1.3 or §4.2.6 of Appendices 30/30A and corresponding Part B submitted under §4.3.14 up to 2 June 2000 and the §4.1.12 or §4.2.16 of Appendices 30/30A shall be subject to a charge in accordance with *decides* 2 above.

7bis That there will be no cost-recovery charges for any submission under §6.17 of Article 6 of Appendix 30B where the associated submission under §6.1 of that Article was received prior to 17 November 2007.

8. That the Annex (Schedule of processing charges) to this decision should be reviewed periodically by the Council.
9. That the payment of charges shall be made on the basis of an invoice issued upon receipt of the filing by the Radiocommunication Bureau and sent to the notifying administration or, at the request of that administration, to the satellite network operator in question within a period of a maximum of 6 months after issue of the invoice.
10. That any subsequent cancellation received by the Radiocommunication Bureau within 15 days of the date of receipt of the filing shall remove the obligation to pay the fee.
11. That publication of special sections for the amateur-satellite service, the notification for recording of frequency assignments for earth stations, for the conversion of an allotment into an assignment in accordance with the procedure of

[5] A submission of filings under Article 4 of Appendix 30 and Appendix 30A in the Regions 1 and 3 Plans, referring to a single orbital position with the same satellite name and received on the same date shall be considered as one "satellite network" filing for the purpose of free entitlement.

former Section I of Article 6 of Appendix 30B, the addition of a new allotment to the plan for a new Member State of the Union in accordance with the procedure of Article 7 of Appendix 30B and submissions under resolves 3 and 4 of Resolution 555 (WRC-12) shall be exempt from any charges.

12. That the date of entry into force of Decision 482 (modified 2013) shall be 1 July 2013.
13. That the provisions of this decision need to be revised when further data from time recording are available,

*Recommends*

That should Council[6] revise the schedule in the Annex, any credits that may arise should be applied by the Bureau to subsequent invoices as requested by administrations.

*Encourages Member States*

To develop domestic policies that will minimize the occurrence of non-payment and consequential revenue loss to ITU.

*Instructs the Director of the Radiocommunication Bureau*

1. To enhance the Radiocommunication Bureau's electronic notice form software (SpaceCap) in order to enable the calculation of the best estimated charges associated with a satellite network filing of any type prior to its submission to ITU.
2. To submit an annual report to the Council on the implementation of this decision, including analyses of:
   (a) The cost of the different steps of the procedures.
   (b) The impact of the electronic submission of information.
   (c) Enhancement in quality of service, including, among others, reduction of the backlog.
   (d) The costs of validating filings and requesting corrective action thereto.
   (e) Difficulties encountered in applying the provisions of this decision.
3. To inform the Member States of any practice used by the Radiocommunication Bureau to implement the provisions of this decision and the rationale for that practice.

## Annex: Schedule of Processing Charges To Be Applied To Satellite Network Filings Received by the Radiocommunication Bureau on or After 1 July 2013

[6] Editorial amendment made by the secretariat.

| Type | | Category | | Flat fee per filing (in CHF) (≥100 units, if applicable) | Start fee per filing (in CHF) (<100 units) | Fee per unit (in CHF) (<100 units) | Cost-recovery unit |
|---|---|---|---|---|---|---|---|
| 1 | Advance publication (A) | A1 | Advance publication of a non-geostationary-satellite network not subject to coordination under Sub-Section **IA** of Article **9**; Advance publication of inter-satellite links of a geostationary-satellite space station communicating with a non-geostationary space station provisionally not subject to coordination in accordance with the Rule of Procedure on No. **11.32**, §6 (MOD RRB04/35)<br>*Note*: Advance publication also includes the application of No. **9.5** (API/B special section) and will not be separately charged | 570 | | Not applicable | |
| 2 | Coordination (C) | C1* | Coordination request for a satellite network in accordance with No. **9.6** along with one or more of Nos. **9.7, 9.7A, 9.7B, 9.11, 9.11A, 9.12, 9.12A, 9.13, 9.14** and **9.21** of Section **II** of Article **9**, **§7.1** of Article **7** of Appendix **30**, **§7.1** of Article **7** of Appendix **30A**, Resolution **33** (Rev. WRC-03) and Resolution **539** (Rev. WRC-03) | 20,560 | 5,560 | 150 | Product of the number of frequency assignments, number of classes of station and the number of emissions, summed up for all frequency assignment groups |
| | | C2* | *Note*: Coordination also includes the application of Sub-Section **IB** | 24,620 | 9,620 | | |
| | | C3* | of Article 9, Nos. **9.5D**, **9.53A** (CR/D special section) and **9.41/9.42** and will not be separately charged | 33,467 | 18,467 | | |
| 3 | Notification (N)[a] | N1*[d] | Notification for recording in the MIFR of frequency assignments to a satellite network subject to coordination under Section **II** of Article **9** (with the exception of non-geostationary-satellite network subject to No. **9.21** only) | 30,910 | 15,910 | | |
| | | N2* | *Note*: Notification also includes the application of Resolutions **4** and **49**, Nos. | 57,920 | 42,920 | | |
| | | N3* | **11.32A** (see footnote a), **11.41**, **11.47**, **11.49**, Sub-section IID of Article **9**, Sections 1 and 2 of Article **13**, Article **14** and will not be separately charged | 57,920 | 42,920 | | |
| | | N4 | Notification for recording in the MIFR of frequency assignments to a non-geostationary-satellite network not subject to coordination under Section **II** of Article **9,** or subject to No. **9.21** only. | 7,030 | | Not applicable | |

| | | | | | |
|---|---|---|---|---|---|
| 4 | Plans (P) | P1 | Part A Special Section for a proposed new or modified assignment in the Regions 1 and 3 List or feeder-link Lists of additional uses under §**4.1.5** or proposed modification to the Region 2 Plans under §**4.2.8** of Appendices **30** or **30A**; or Part B Special Section for a proposed new or modified assignment in the Regions 1 and 3 List or feeder-link Lists of additional uses under §**4.1.15** (except Part B special section related to the application of Resolution **548** (WRC-03)) or proposed modification to the Region 2 Plans under **4.2.19** of Appendices **30** or **30A**[b] | 28,870 | Not applicable |
| | | P2[d] | Notification for recording in the MIFR of frequency assignments to space stations in the broadcasting-satellite service and its associated feeder-link in Regions 1 and 3 or Region 2 under Article **5** of Appendices **30** or **30A**[b] | 11,550 | |
| | | P3 | Coordination request in accordance with Article **2A** of Appendices **30** and **30A** | 12,000 | |
| | | P4 | Request for the conversion of an allotment into an assignment with modification which is beyond the envelop characteristics of the initial allotment, or for the introduction of an additional system, or for the modification of an assignment in the List in accordance with §6.1 of Article 6 of Appendix **30B**; or request for inclusion of assignments into the List for converted allotment with modification which is beyond the envelop characteristics of the initial allotment, or for an additional system or for modified assignments in the List in accordance with §6.17 of Article 6 of Appendix **30B**[c] | 25,350 | |
| | | P5[d] | Notification for recording in the MIFR of frequency assignments to space stations in the fixed satellite service under Article **8** of Appendix **30B** | 20,280 | |

[a]Fees for Categories N1, N2 and N3 are applicable to the first notification of assignments that also contains a request to apply No. **11.32A**. If the application of No. **11.32A** is not requested, 70% of the indicated fees will apply, with the remaining 30% to be charged to a subsequent request, if any, for application of No. **11.32A**

[b]Under this category, taking account that a filing for the broadcasting-satellite service and its associated feeder link in Region 2 includes both the downlink (AP30) and the feeder link (AP30A), which are examined and published together, the total fee application to such filing shall be twice the fee indicated in the column "Flat fee per filing"

[c]Fees for a request in accordance with §6.17 of Article 6 of Appendix **30B** also contains a possible subsequent request (resubmission) in accordance with §6.25. A request in accordance with §6.17 of Article 6 of Appendix **30B** for a submission treated as that under §6.1 in accordance with §7.7 of Article 7 shall not be charged

[d]For cases of consolidation of frequency assignments in the MIFR of different GSO networks submitted by an administration (or an administration acting on behalf of a group of named administrations) under Article 11 of the Radio Regulations, category N1 shall apply, for cases submitted under Appendices 30 or 30A, category P2 shall apply, and for cases submitted under Appendix 30B, category P5 shall apply

***Definition of category for coordination (C) and notification (N)**

The category for coordination (C1, C2, C3) and for notification (N1, N2, N3) is related to the number of forms of coordination applicable to a particular satellite network coordination request or notification submission, as follows:

- C1 and N1 correspond to a satellite network filing referring to only one cost-recovery form of coordination (A, B, C, D, E or F). Both categories also include cases for which no form of coordination applies as a result of unfavourable finding under No. 11.31 of the Radio Regulations for all frequency assignments of the submitted filing, or cases including frequency assignments published for information only.
- C2 and N2 correspond to a satellite network filing referring to any two or three cost-recovery forms of coordination amongst A, B, C, D, E or F.
- C3 and N3 correspond to a satellite network filing referring to any four or more cost-recovery forms of coordination amongst A, B, C, D, E or F.

| Cost-recovery form of coordination | Individual radio regulations forms of coordination |
|---|---|
| A | No. 9.7, RS33.3 |
| B | AP30 7.1, AP30A 7.1 |
| C | No. 9.11, RS33 2.1, RS539 |
| D | Nos 9.7B, 9.11A, 9.12, 9.12A, 9.13, 9.14 |
| E | No. 9.7A[a] |
| F | No. 9.21 |

[a]Cost recovery for category C1 only. See also *decides* 11

MIX
Papier aus verantwortungsvollen Quellen
Paper from responsible sources
FSC® C105338

If you have any concerns about our products,
you can contact us on
**ProductSafety@springernature.com**

In case Publisher is established outside the EU,
the EU authorized representative is:
**Springer Nature Customer Service Center GmbH**
**Europaplatz 3, 69115 Heidelberg, Germany**

Printed by Libri Plureos GmbH
in Hamburg, Germany